JEFFREY

Coal and Ashes
Handling Machinery

*Power House
Division*

Catalog No. 385

THE JEFFREY MANUFACTURING CO.
COLUMBUS, OHIO, U. S. A.

——SALES OFFICES——

NEW YORK, 30 Church Street
CHICAGO, McCormick Building
PITTSBURGH, 620 Second Avenue
PHILADELPHIA, Real Estate Tr. Bldg.
BOSTON, 141 Milk Street

CLEVELAND, 1128 Guardian Bldg.
SCRANTON, Union Nat'l Bank Bldg.
CINCINNATI, 62 Plum Street
ST. LOUIS, 606 Pontiac Building
CHARLESTON,W.Va.,914 Kanawha St.

DETROIT, Book Building
DENVER, 1751 Wazee Street
MONTREAL, Power Building
MILWAUKEE, M. & M. Building
LOS ANGELES, H. W. Hellman Bldg.

Copyright 1924 by The Jeffrey Mfg. Co.

3-24-10M-M

Why Leaders of Industry Prefer Jeffrey Equipment

IN large power plants there must be no interruption of service, no danger of shutdowns, no inability to handle peak loads.

This is empathically true of the boiler house, where even the slightest failure, as for example the breakdown of Conveying Apparatus, is likely to cause a serious loss.

For many years, Jeffrey Coal and Ashes Handling Equipment has been safe-guarding the country's greatest power plants and helping build their modern high standards of operation.

And for nearly half a century Jeffrey has been the leading source of supply for coal mining and coal handling equipment. It is quite natural, therefore, that Industrial Leaders in the power plant field should turn to Jeffrey for the solution of their coal and ashes handling problems.

The results of this long experience have been combined between the covers of this book and are intended to help the Engineer in the selection of the proper type of equipment for crushing, handling or storing of coal to meet his particular condition or requirement.

Typical layouts giving general dimensions accompany the tables of specifications for various equipments. The data shown permits of Jeffrey products being laid in on plans, thus eliminating possible delay due to requesting information from the home office.

Our Engineering experts are at your service and will gladly assist you in the selection of Coal and Ashes Handling Equipment applicable to your requirements.

Pivoted Bucket Carrier

Section

1

Upper Run of the Jeffrey Pivoted Bucket Carrier over bunkers

The Highest Type of Coal and Ashes Handling Equipment

YEARS of broad experience and study of Boiler House operation have enabled Jeffrey Engineers to produce in the Pivoted Bucket Carrier, the highest, most efficient and dependable type of Coal and Ash Handling Equipment. The design, workmanship and special methods used in the construction of this Carrier give assurance of proper performance and long life.

Built in and around the complete sense of the terms Reliability and Long Service, the Jeffrey Carrier insures reliability in its action at all times, whether the installation outline be intricate, the surrounding conditions crude, or the attendant labor not of a high class. Reliability First has been considered paramount coupled with that material strength and hardness of wearing parts throughout, which means for long and satisfactory service from the installation as a whole.

The Jeffrey Carrier has a working principle which ordinarily combines in the one carrier the work of at least one elevator and two conveyors, in that it handles both coal and ashes. Pages 18 and 19 show a typical installation of the Carrier in detail.

No Mechanical Loader Required with the Jeffrey Carrier

Along the horizontal and upon an incline the Jeffrey Carrier, by reason of the overlapping lips of its buckets, is virtually a continuous apron with depressions in its surface formed by the openings of the buckets. The Carrier therefore does not require an automatic loader to deposit a separate load into each bucket, but on the contrary permits a continuous stream of materials to be discharged onto the Conveyor.

The economical installation of the Jeffrey Pivoted Bucket Carrier is usually a minimum capacity of 50 tons per hour of coal with a maximum capacity of about 175 tons per hour.

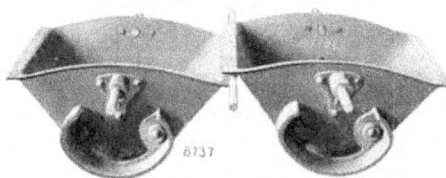

4

Typical Installations

THE economy and efficient performance of Jeffrey Coal and Ashes Handling Equipment is very evident at the Ford River Rouge Plant. Here, in the pulverizer building, coal is ground into boiler fuel and then conveyed through the pulverizer building up to the tower by a Jeffrey Pivoted Bucket Carrier on to the cross conveyors into the boiler room.

Typical Installations

OPERATING on an average of fourteen hours every day in the year, handling both ashes and crushed coal at the power plant of the Transue Williams Steel Forging Corporation, Alliance, Ohio. This 18″ x 24″ Jeffrey Carrier is 48 feet vertical by 165 feet horizontal and requires about seven horsepower at full capacity. A Jeffrey 30″ x 30″ Single Roll Crusher is used under the Track Hopper. Note Ash Bin with unloader in the foreground.

Typical Installations

The gradual moving of the Traveling Tripper will make a continuous ridge of the pile of coal shown in the foreground and then completely fill the bins from edge to edge.

On the left is the completely filled Steel Apron Conveyor from Crusher to Pivoted Bucket Carrier at the Transue Williams Plant. The Apron is 30 inches wide and ends 4 inches high.

Boiler room of the Transue Williams Plant. Above these cemented bunkers with their easily controlled valves and flexible spouts, the Jeffrey 18" x 24" Carrier is in almost constant service.

Typical Installations

THE Solar Refining Co., Lima, Ohio. A power plant where a Jeffrey 24″ x 24″ Carrier is handling about 70 tons of crushed coal per hour. The vertical lift is 66 feet while the horizontal run is 107 feet. About 8 to 10 horsepower is required for the full capacity of this carrier. A Jeffrey 30″ x 30″ Single Roll Crusher fed by a steel apron from beneath a track hopper insures a uniform product to the Carrier for the stokers.

Typical Installations

Upper Run of the Carrier at the Solar Plant. Note the readily adjustable Traveling Tripper. Its dumping cams are chilled hard as flint for long service.

In the basement of the Solar Refining Company's Power Plant, showing the lower run of the Jeffrey Carrier with its arched heavy steel skirts located directly beneath the Single Roll Crusher.

9

Typical Installations

POWER Plant of the Sun Oil Co. at Marcus Hook near Philadelphia, Pa. The vertical lift of this 18″ x 24″ Jeffrey Carrier is 56 feet while the horizontal run is 82 feet. Only 5 to 6 horsepower is required for its full capacity rating of 50 tons per hour of crushed coal or ashes.

A track hopper with steel apron loader to a 24″ x 24″ Jeffrey Crusher feeds the Carrier.

Typical Installations

Inside the Boiler House of the Sun Oil Co. Note the Traveling Tripper in the upper run and the down casing in the background.

Interior of the Boiler Room of the Plant illustrated on the opposite page, showing how the coal is distributed from overhead bunkers, kept filled with a Jeffrey Pivoted Bucket, to stokers by means of chutes.

A Steel Apron Conveyor transfers coal from track hopper to Coal Crusher from which the Pivoted Bucket carries it to overhead bunkers.

This Pivoted Bucket is also used to deliver the ashes to storage bin.

Typical Installations

11440

AT the plant of the Oliver Chilled Plow Company at South Bend, Indiana, a complete Jeffrey
Coal and Ashes Handling Outfit is used, the principal units of which include Track Hopper,
Plate Feeder, Single Roll Crusher, and Pivoted Bucket Carrier for both coal and ashes handling.

Typical Installations

Upper run of the Jeffrey Pivoted Bucket Carrier in operation in the plant shown on opposite page. It has a capacity of 50 tons of crushed coal per hour.

The lower horizontal run of conveyor shown above. It also handles the ashes. The steel skirt boards protect the chain from grit.

Jeffrey Track Hopper, Plate Feeder, Single Roll Crusher and Apron Conveyor are also a part of the Oliver Equipment.

A corner in the basement of the Oliver plant showing the Apron Conveyor delivering crushed coal to Pivoted Bucket Carrier.

Typical Installations

AT the power house of the Hershey Chocolate Co., Hershey, Pa., coal is received on the concrete siding, fed to Jeffrey Single Roll Crusher and flows by gravity to a Jeffrey Pivoted Bucket Carrier which delivers crushed coal to storage pile or into bunkers above the boilers.

The same Conveyor reclaims the coal from the storage pile, and also handles the ashes on the lower run. The above diagram shows the outline of the conveyor system.

Typical Installations

12849

Above is shown the Pivoted Bucket Carrier for handling coal from outside storage at the Hershey Chocolate Plant. Ashes are also delivered by this carrier to railroad cars as will be noted by the chute arrangement over tracks.

The upper run of this Carrier over bunkers is shown in the left hand illustration.

Lower run of the above installation, where coal is received from outside storage. Ashes are also received through similar chutes under boilers and are carried to railroad cars.

12853-1

Typical Installations

THE coal handling system at the Armour Power House No. 3 consists of Double Track Hoppers, Apron Feeder, Crusher and Pivoted Bucket Carrier. The capacity of the entire system is about 67 tons per hour of crushed stoker size coal. The Apron Feeder consists of steel beaded flights mounted upon two strands of No. 809 steel thimble roller chain. The Pivoted Bucket Carrier is the popular 24" x 24" size of bucket mounted on the double bushed type of chain. This system has a lift of approximately 80 feet with a horizontal run of 83 feet.

Typical Installations

HOUSATANIC Power Co., Waterbury, Conn. At this plant the Jeffrey 24″ x 24″ Carrier is handling both Coal and Ashes. Installed by R. H. Beaumont Co., Contracting Engineers. Coal is unloaded direct from hopper bottom cars to the Carrier through a double track hopper as indicated in the sketch to the right—or by a grab bucket locomotive crane from yard storage to the Elevated Hopper shown in illustration above.

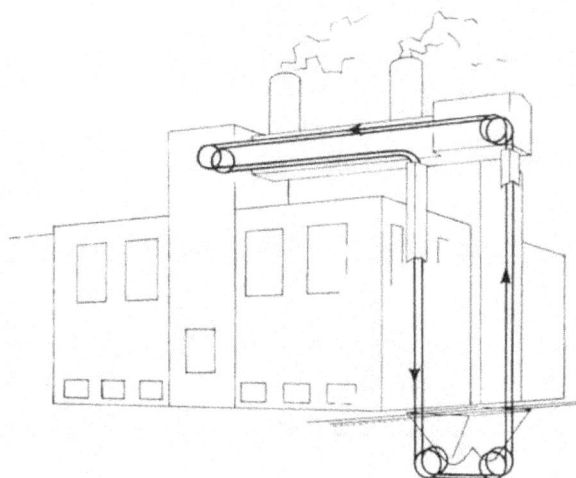

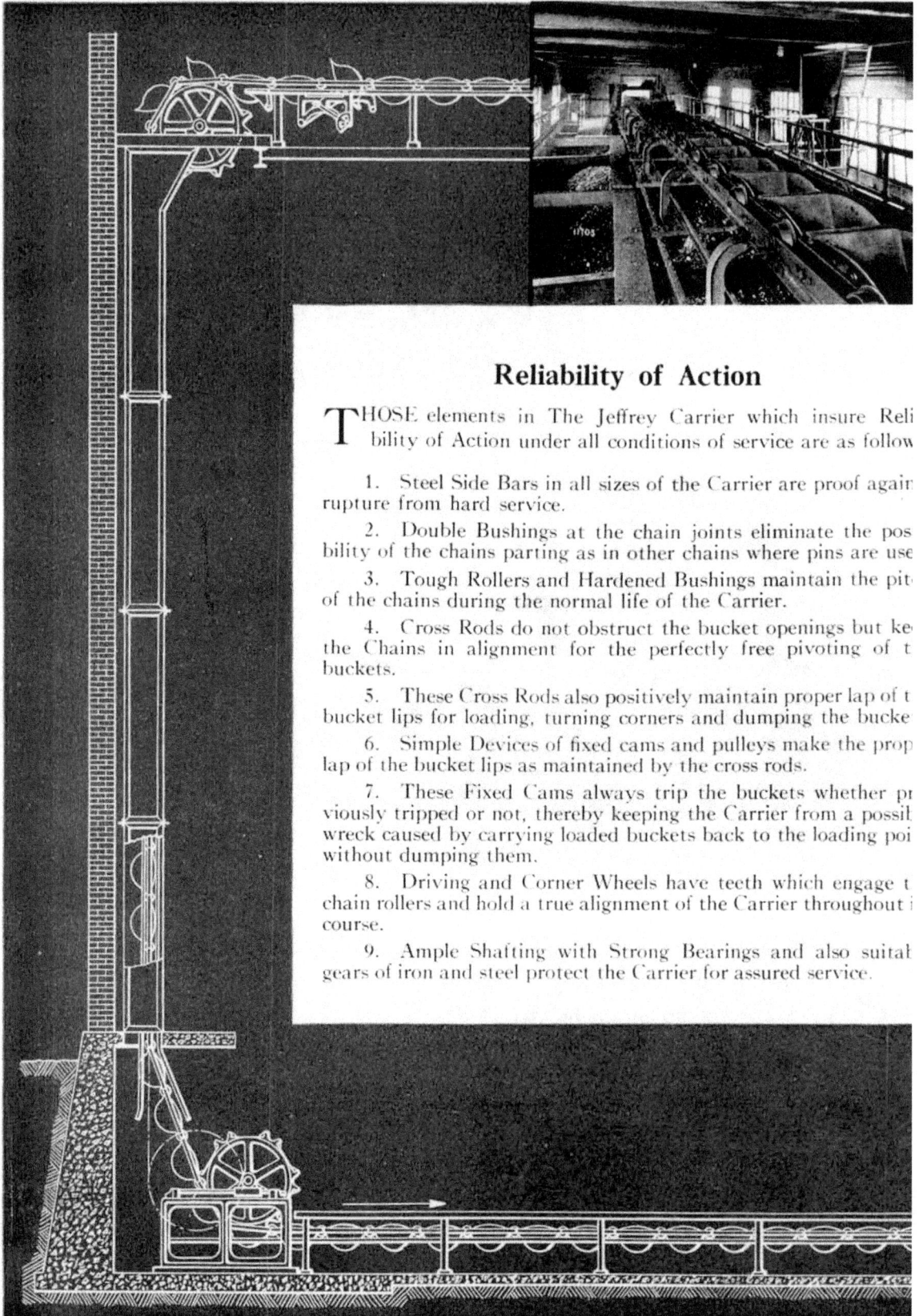

Reliability of Action

THOSE elements in The Jeffrey Carrier which insure Reli
bility of Action under all conditions of service are as follow

1.   Steel Side Bars in all sizes of the Carrier are proof again
rupture from hard service.

2.   Double Bushings at the chain joints eliminate the pos
bility of the chains parting as in other chains where pins are use

3.   Tough Rollers and Hardened Bushings maintain the pit
of the chains during the normal life of the Carrier.

4.   Cross Rods do not obstruct the bucket openings but ke
the Chains in alignment for the perfectly free pivoting of t
buckets.

5.   These Cross Rods also positively maintain proper lap of t
bucket lips for loading, turning corners and dumping the bucke

6.   Simple Devices of fixed cams and pulleys make the prop
lap of the bucket lips as maintained by the cross rods.

7.   These Fixed Cams always trip the buckets whether pr
viously tripped or not, thereby keeping the Carrier from a possil
wreck caused by carrying loaded buckets back to the loading poi
without dumping them.

8.   Driving and Corner Wheels have teeth which engage t
chain rollers and hold a true alignment of the Carrier throughout i
course.

9.   Ample Shafting with Strong Bearings and also suitab
gears of iron and steel protect the Carrier for assured service.

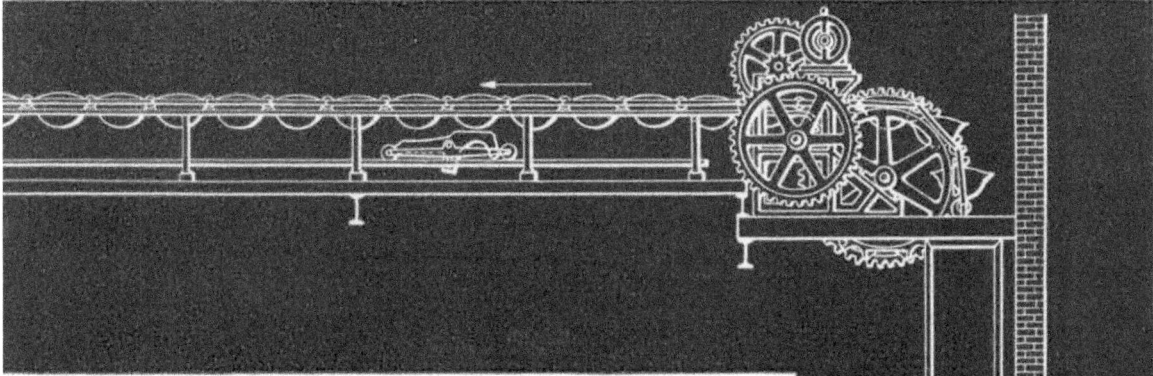

Long Service Qualities

IN distinction to those elements upon the opposite page covering Reliability, below is given those qualities which insure Long Service.

1. Steel Side Bars of high carbon and hard to bend, but very tough. Their thickness gives ample bearing for the Double Bushings.

2. Double Bushings, glass hard and tough, are tight fit in side bars. Wear of chain joints is between bushings, not in side bars.

3. Large White Iron Rollers, tough and hard on tread and in the bore, twice as hard as good gray iron.

4. Steel Cross Rods connecting the chains receive no perceptible wear. They simply touch the bucket lips.

5. Overlapping Lip Buckets are either of tough malleable iron for coal and ashes, and gray steel for cement and hot materials, or cast ends with renewable steel bottoms depending upon the size.

6. Tripping Cams on Buckets are glass hard on their wearing surfaces, and when not cast integral are hot riveted to the bucket ends.

7. Steel Pivot Pins of the buckets are heat treated to glass hardness and bear in hard white iron pivot blocks in the chains.

8. Stationary and Traveling Trippers have deep chilled wearing surfaces and also can be readily lowered when not in service.

9. Driving gears are in keeping with the other high qualities of the Carrier for Long Service. Gear Guards protect them.

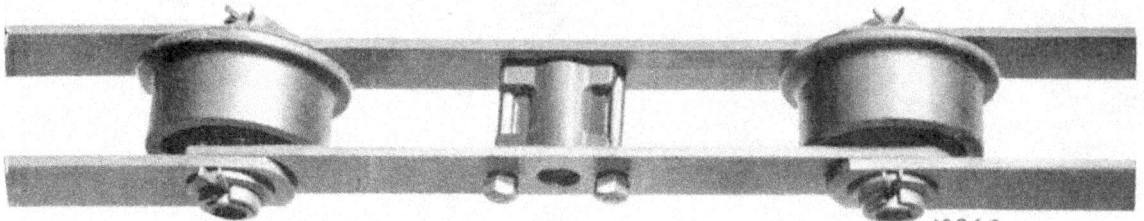

13969

(Patented)

THE life of any Carrier in rigidity and in proper action on its driving wheels depends almost entirely upon the life of its chain joints. The Jeffrey Chain therefore is made of tough steel side bars with hardened double bushings. These two bushings are assembled one within the other, the outer one fixed to the inside bars and the inner bushing fixed to the outside bars.

The outer bushings serve as bearings for the chain rollers while the inner bushings act as chain pins and receive the carrier cross rods. It is to these cross rods that the chains are readily assembled by means of malleable washers and large cotter pins on each side of the chain joints. Lugs in the steel side bars are locked tight to notches in both ends of the bushings. Thus all wear is confined to the bushings which may be replaced after long service and a practically new chain obtained at very little cost.

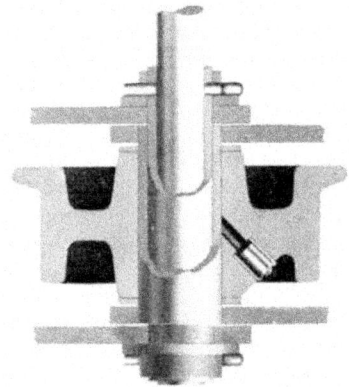

The chain roller, with its thick tread and heavy flange, has been designed to withstand all the shrinkage strains incident to being cast of an exceedingly tough and very hard, white iron. In fact, the roller is so hard that it resists the action of high-grade machine tools, and necessitates the bore be ground on a special grinding machine. The Rollers are equipped with a well known "High Pressure" Lubrication System.

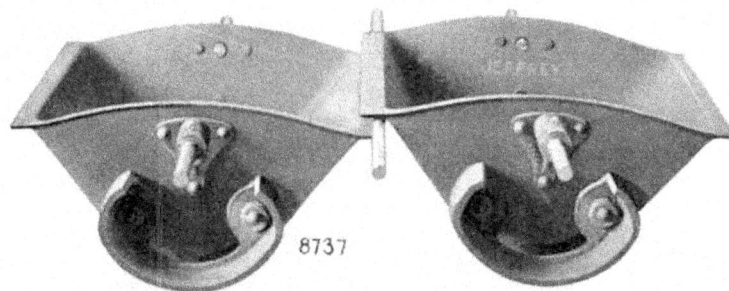

8737

The shape of the Jeffrey Pivoted Bucket is the result of numerous tests to obtain not only the greatest capacity possible, but also a complete discharge of the material handled with the least dumping effort or shock to the Carrier and the least wear of the bucket cams and pivots. The cross-rods prevent rocking of the buckets and maintain a true alignment of the steel chains for a perfect pivoting action of the buckets. Besides being cast from high-grade materials, the Jeffrey Bucket is reinforced on every edge and corner where long experience has dictated to insure the longest life possible. The Buckets are suspended at their ends upon hardened steel trunnion pins pivoted in white iron blocks bolted between the side bars and midway between joints of the chain. The replacing of buckets is a simple operation, and in no way affects the chains.

Stationary Tripper

The Stationary Tripper is used primarily where there is but one point of discharge, whereas the Traveling Tripper is used where a long storage bin or space is to be completely and uniformly filled.

Showing the Stationary Tripper in the act of tripping the Buckets over the Ashes Chute.

Traveling Tripper

Tripping Cams are readily lowered by rack and pinions so that Tripper may be quickly shifted over a long space when the Carrier is idle or when a Stationary Tripper is in service.

Traveling Tripper

The Traveling Tripper is moved by means of a 3/8" steel wire rope having several wraps around a drum, operated by a large hand wheel, which insures rapid shifting of the tripper from place to place.

Left hand view shows the Traveling Tripper disengaged, while the right hand illustration clearly shows method of tripping buckets.

Both Traveling and Stationary Trippers tilt the buckets so that the discharge bottom is somewhat in excess of 55 degrees to the horizontal.

Pivoted Bucket Carrier =========================== ===== JEFFREY=

Table of Sizes and Capacities of Jeffrey Pivoted Bucket Carriers

Size Carrier				Tons per Hr. Coal–50 lb. cu. ft. Speed 50 F. P. M. §	Power Factors—Materials Weighing					
Bucket †		Chain			50 lbs. per cu. ft.		100 lbs. per cu. ft.		150 lbs. per cu. ft.	
Width	Length	Pitch	Roller		A	B	A	B	A	B
					For Lift L	For Run R	For Lift L	For Run R	For Lift L	For Run R
16″	18″	18″	5″	30	.043	.026	.086	.032	.129	.038
18″	24″	24″	6″	50	.064	.030	.128	.039	.192	.048
24″	24″	24″	6″	70	.085	.035	.170	.047	.255	.060
30″	24″	24″	6″	85	.106	.044	.212	.060	.320	.076
30″	30″	30″	7″	125	.136	.053	.272	.072	.408	.092
36″	30″	30″	7″	150	.164	.058	.328	.080	.490	.106

† The three largest buckets are formed of cast ends with their chilled dumping cams integral and with steel plate bodies riveted to the ends, while the smaller sizes are made of malleable iron with chilled dumping cams riveted on.

§ Maximum Speed 60 feet per minute

Figuring the Horsepower Required for Jeffrey Carriers

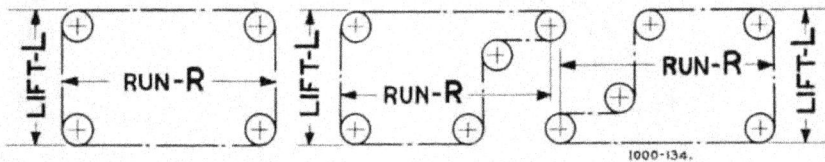

AS the empty vertical strands of the Carrier balance each other, the power to move them is only that required to elevate the load in the up-going strand; while the power to move the horizontal runs is that required to overcome the rolling friction of both the horizontal runs of the Carrier including the loads in them. Therefore, we have in the Table above two Power Factors for a Carrier, "A" for each foot of Vertical Lift "L", and "B" for each foot of Horizontal Run "R"; whereby the power required at the head shaft for 50, 100, and 150 pound per cubic foot materials in a Carrier traveling at 50 ft. per minute speed may be readily figured from the formula:

$$\text{HORSE POWER AT HEAD SHAFT} = (A \times L) + (B \times R).$$

To the power thus obtained add $33\frac{1}{3}\%$ for losses through the gears to obtain the size of the Motor.

For Example: The Motor required to operate an 18 in. x 24 in. Carrier of 75 feet Vertical LIFT and 100 feet Horizontal RUN handling Coal at 50 lbs. per cubic foot and traveling 50 feet per minute is:

$(.064 \times 75) + (.030 \times 100) = 7.8$ Horse Power, which, when increased $33\frac{1}{3}\% = 10.4$ Horse Power at the Motor.

Placing the Jeffrey Carrier Into Your Building Plans

Upon the pages following are blue prints of typical erection drawings for six sizes of the Jeffrey Carrier. These prints give dimensions which are vital to the proper installation of the Carriers.

Thus a Jeffrey Carrier can be definitely embodied in your building plans at the beginning, and time and expense saved thereby.

22

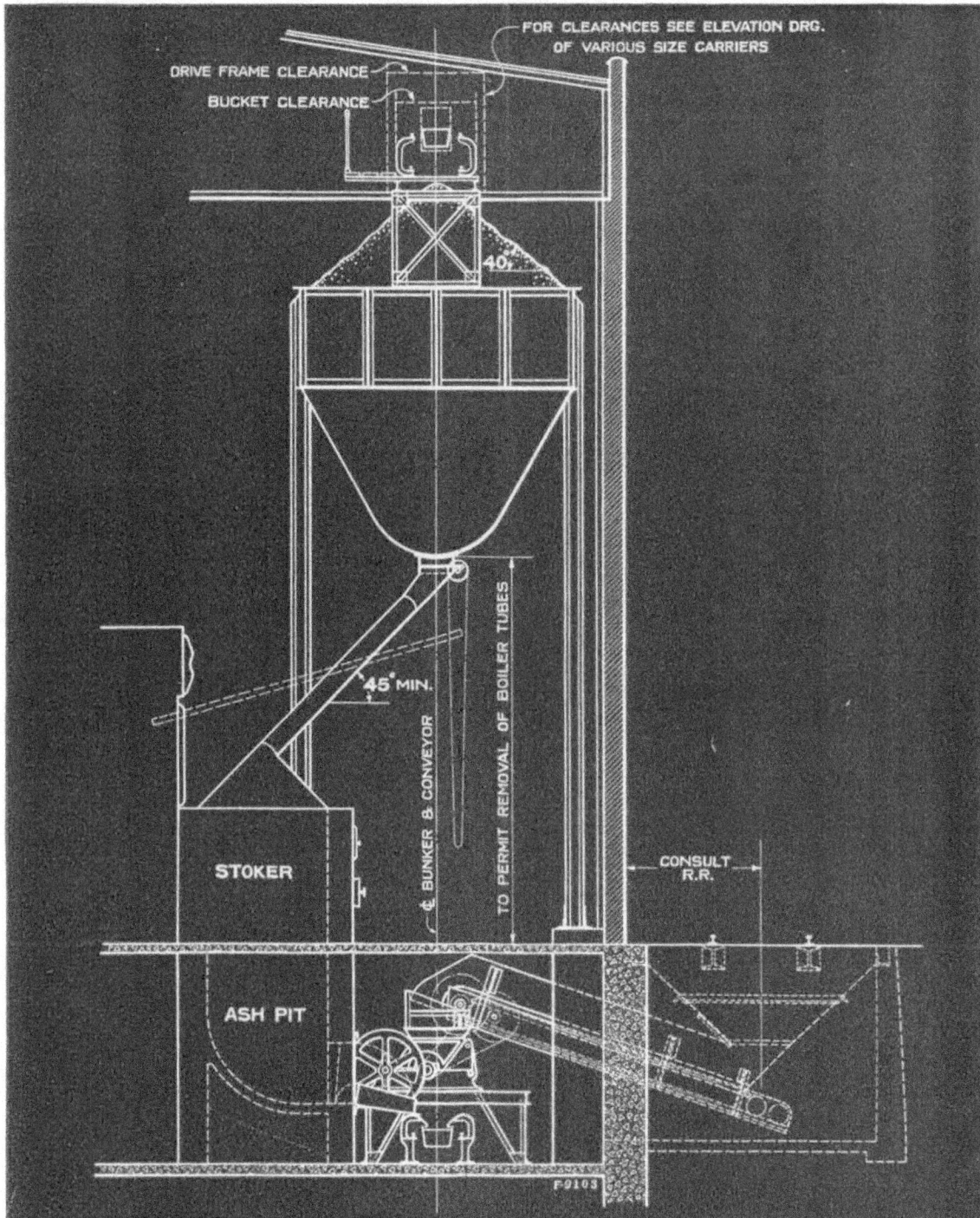

TO assist the architect and engineer to make building plans for the proper reception of The Jeffrey
Carrier, the above blue print, with others on the following pages, have been carefully prepared.
Before using any of the prints, read all notations upon the same.

On the above print note—(1) the 40 degree coning angle of coal in bunker—(2) the minimum
45 degree angle of spouts to stokers—(3) proper clearances below bunkers and in front of stokers.

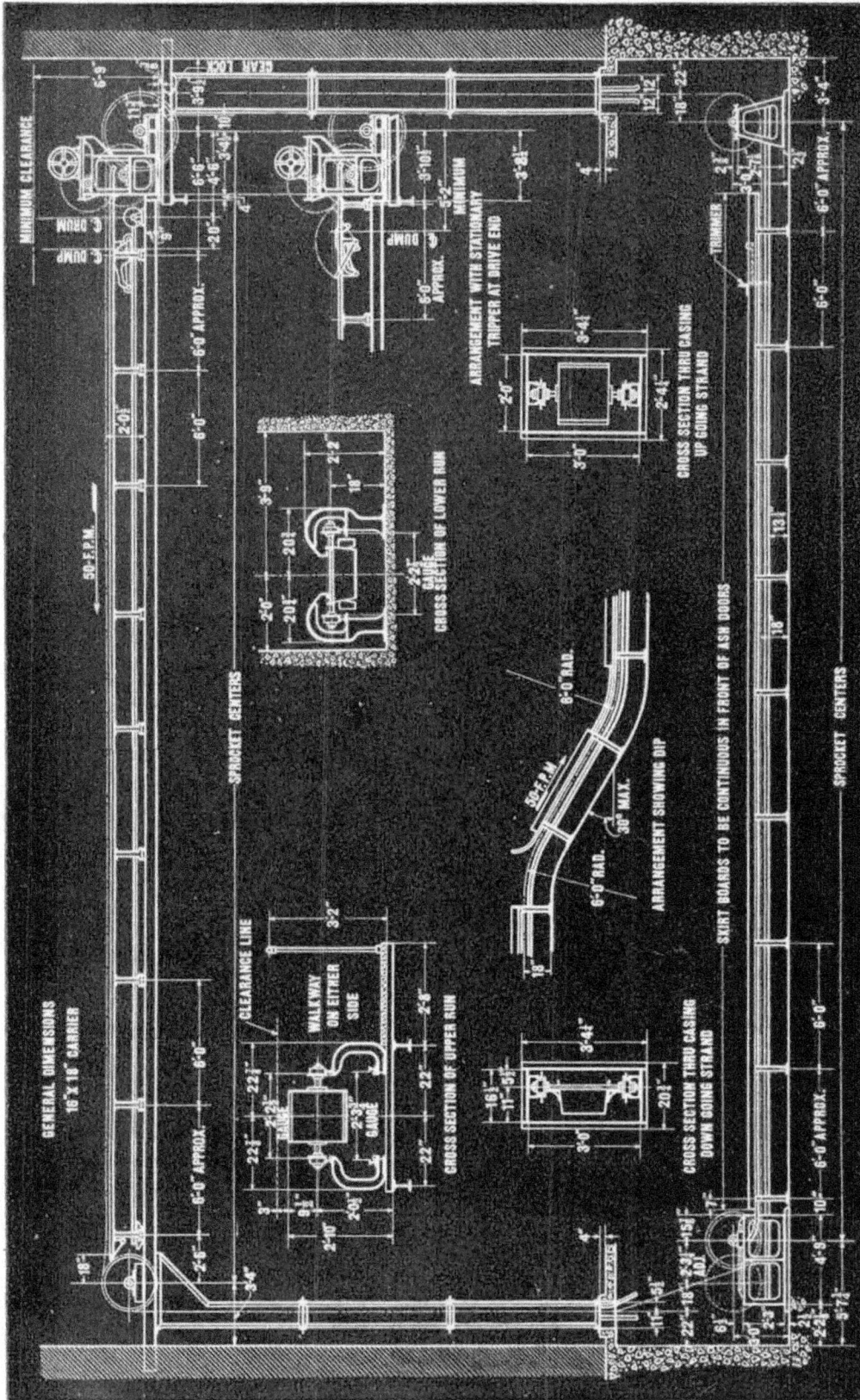

The 16" x 18" Carrier handles 30 tons of coal per hour at 50 feet per minute. Cross channels bolted to upper rail chairs are furnished as part of Standard Equipment for all sizes of Carriers. Where Vertical Lift is less than 15 per cent of Horizontal Run, place the Drive at opposite upper corner from that shown with direction of carrier travel remaining unchanged. Plan Views given on next page.

Plan Views of 16" x 18" Carrier—Upper Run shows the cross channels which are furnished with upper rail chairs. Note extensions for walkway. Lower Run shows the minimum space about the Carrier for proper care and inspection. Where possible increase this clearance along both sides of the Carrier. Anchor bolts in lower corners should be set deep in good cemented work.

The 18″ x 24″ Carrier handles 50 tons of coal per hour at 50 feet per minute. Walkway cross channels are furnished with upper chairs. In "Cross Section of Lower Run," note protection of Carrier chains by rounded loading skirts. Where Vertical Lift is less than 15 per cent of Horizontal Run, place the Drive at opposite upper corner from that shown with direction of carrier travel remaining unchanged. Plan Views given on next page.

Note the anchor bolt plan which gives the location of bolts for fastening the casing to floor. Plan of the Lower Run gives the minimum clearance about the Carrier—but where possible increase this clearance along both sides of the Carrier. Upper Corner supports for all Carriers to be cemented into pilastered end walls or well braced down into bin or building structure. Anchor bolts for Lower Corners to be set deep in good cemented work.

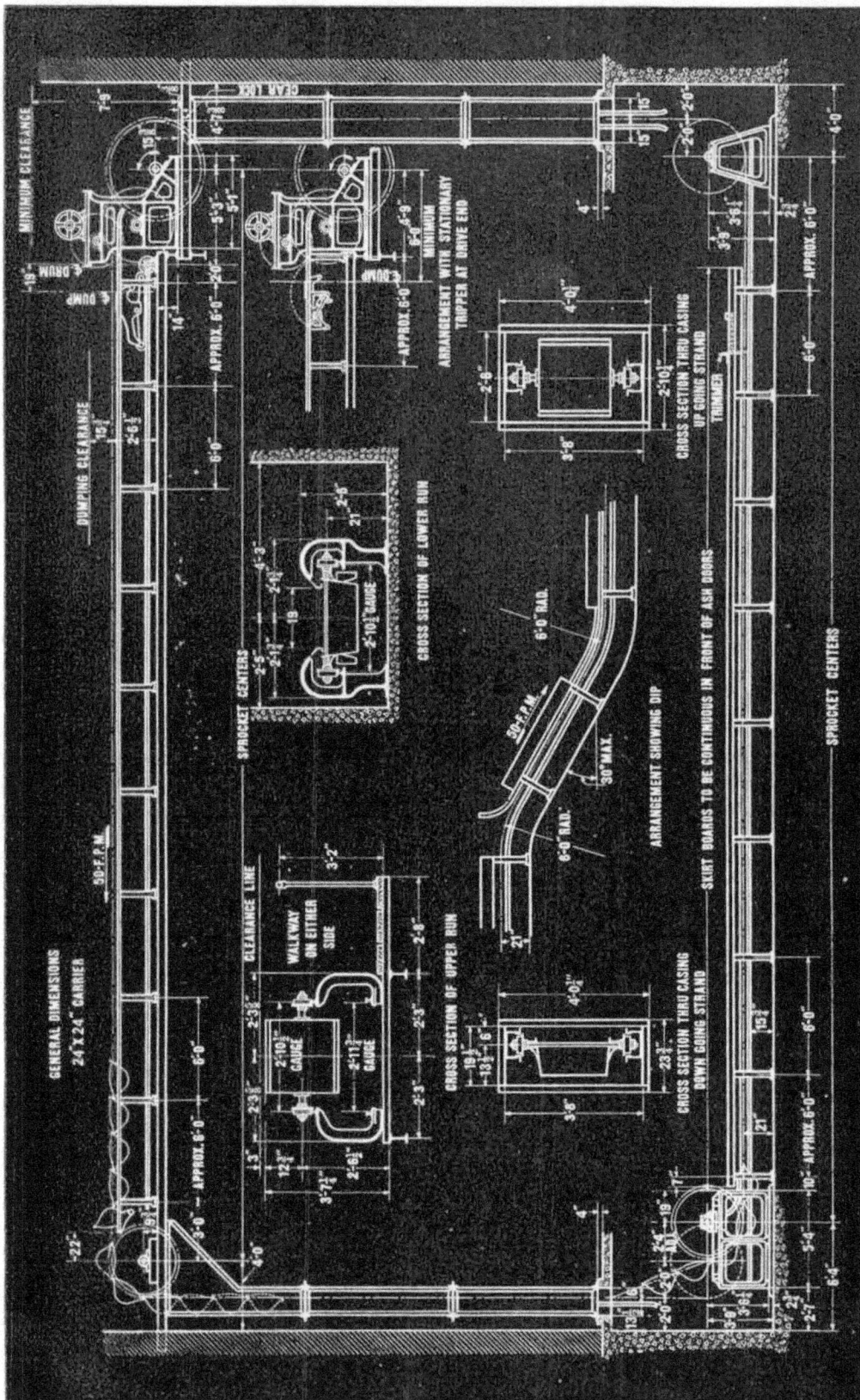

The 24" x 24" Carrier handles 70 tons of coal per hour at 50 feet per minute. Note that foundations under Lower Corner Frames for all sizes of Carriers extend above the general floor level. The closest dumping position of Traveling and Stationary Trippers to the Drive Corners is shown above for 24" x 24" Carrier. Note maximum incline for all Carriers is 30 degrees with 6'-0" minimum radii. Plan Views given on next page.

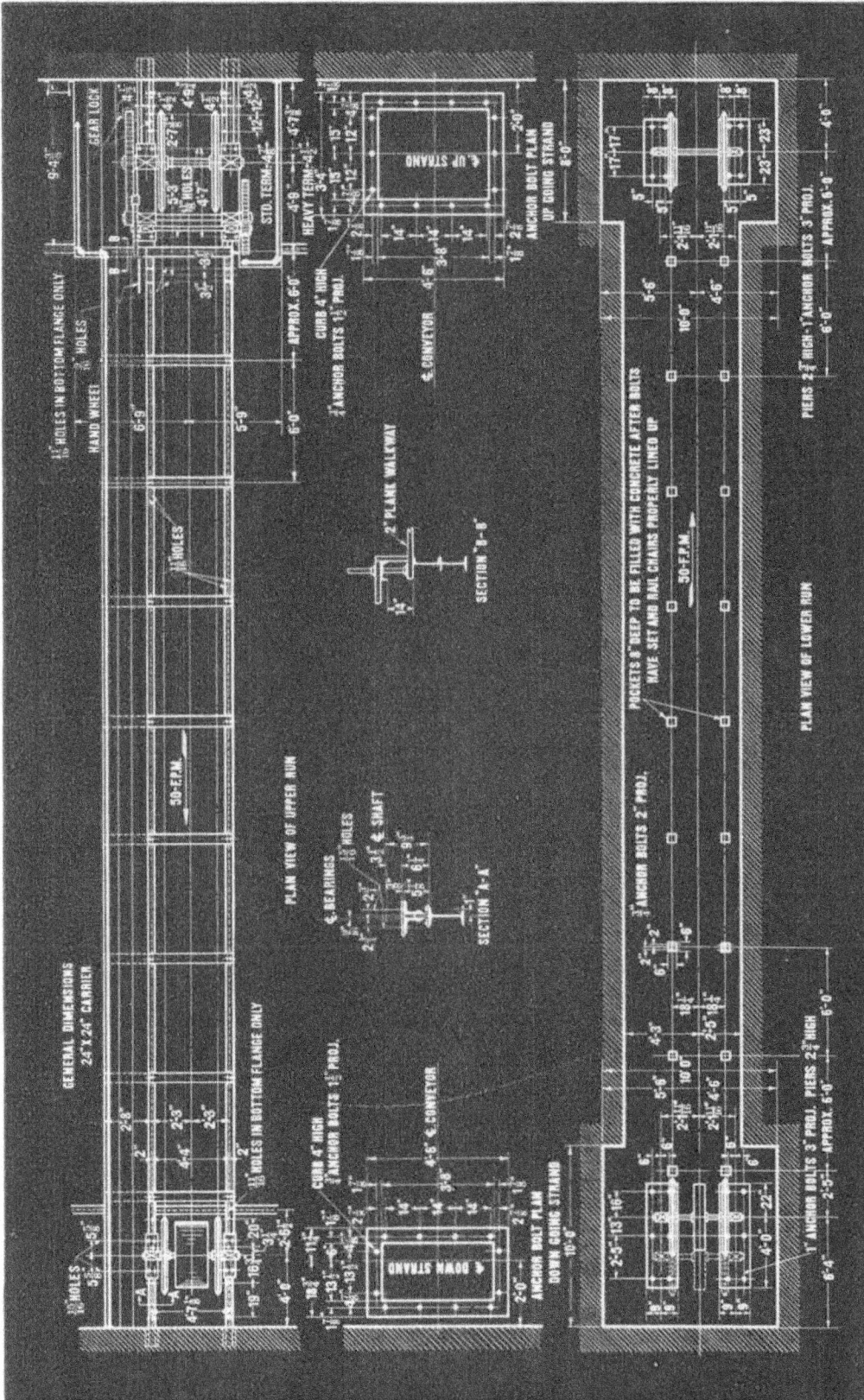

Above Plan of Lower Run gives the minimum clearance about Carrier—but where possible increase this clearance along both sides of the Carrier. Walkway with hand rail should be provided for the proper care and inspection of parts along the Upper Run. Where the Vertical Lift is less than 15 per cent. of the Horizontal Run, place Drive at opposite upper corner from that shown with direction of travel unchanged.

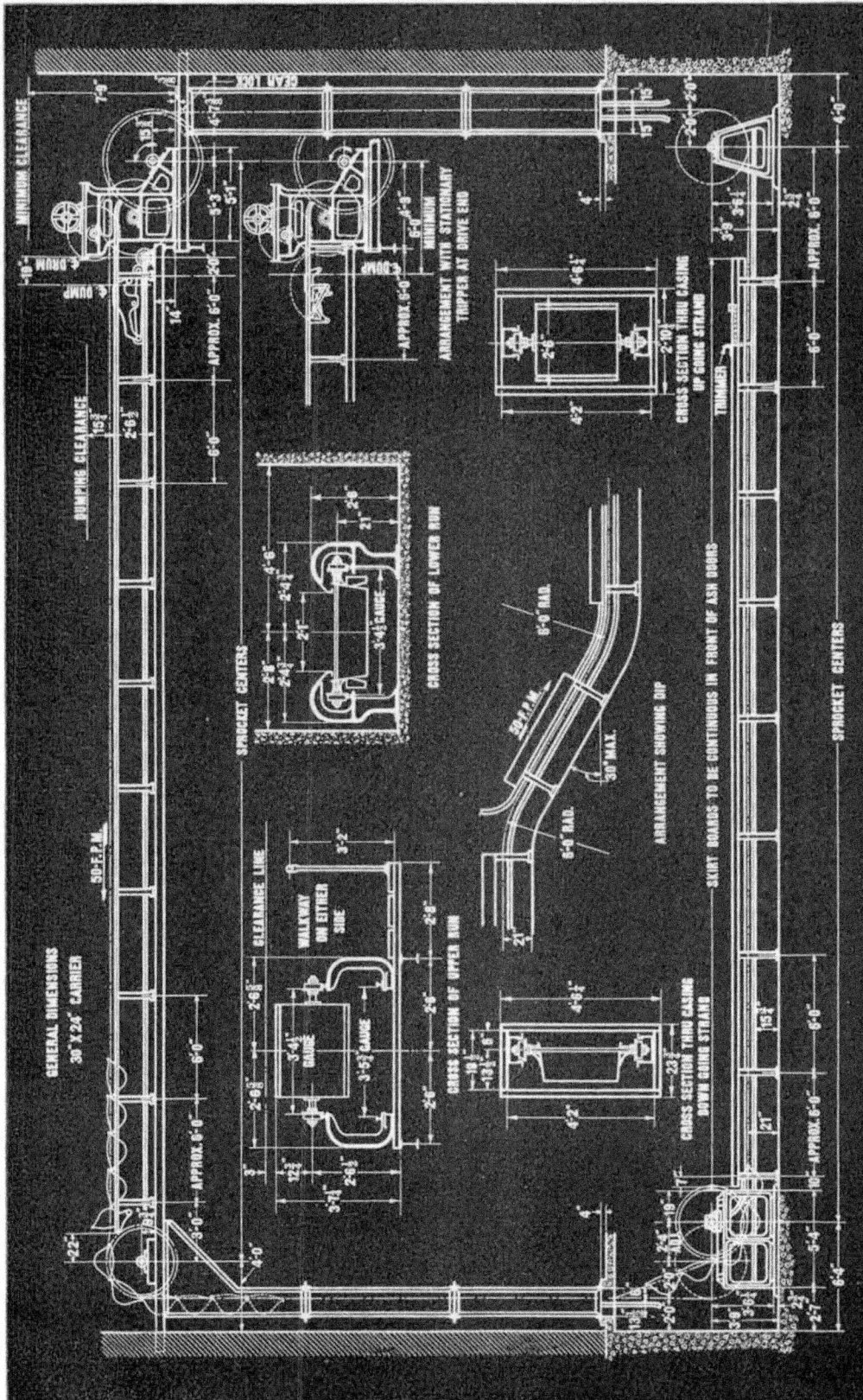

The 30″ x 24″ Carrier handles 85 tons of coal per hour at 50 feet per minute. Walkway cross channels are furnished with upper chairs. Note pads should be provided under the lower corner frames and casings. Maximum incline for all Carriers is 30 degrees with 6′-0″ minimum radii. Plan Views given on next page.

GENERAL DIMENSIONS
30" X 24" CARRIER

PLAN VIEW OF UPPER RUN

30-F.P.M.

ANCHOR BOLT PLAN
UP GOING STRAND

ANCHOR BOLT PLAN
DOWN GOING STRAND

₵ BEARINGS

₵ SHAFT

SECTION A-A

2" PLANK WALKWAY

SECTION B-B

₵ CONVEYOR

CURB 4" HIGH
⅞" ANCHOR BOLTS 1½" PROJ.

PLAN VIEW OF LOWER RUN

POCKETS 8" DEEP TO BE FILLED WITH CONCRETE AFTER BOLTS
HAVE SET AND RAIL CHAIRS PROPERLY LINED UP

50-F.P.M.

PIERS 2¼" HIGH-1" ANCHOR BOLTS 3" PROJ.

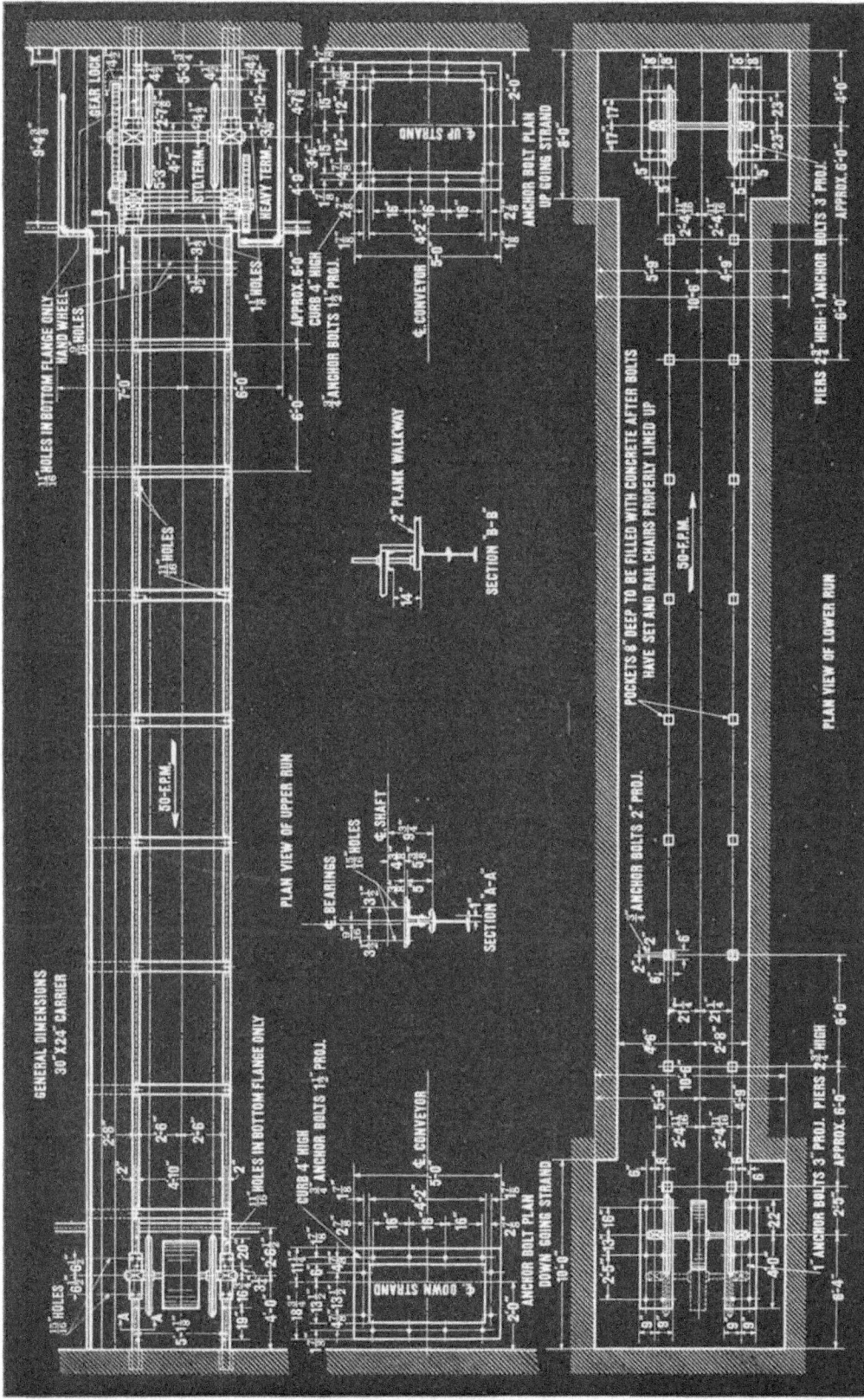

The above plan of Lower Run gives the minimum clearance about the Carrier—but where possible increase this clearance along both sides of the Carrier. Extensions should be provided for Walkway as shown around drive corner to allow for proper care and inspection of driving mechanism.

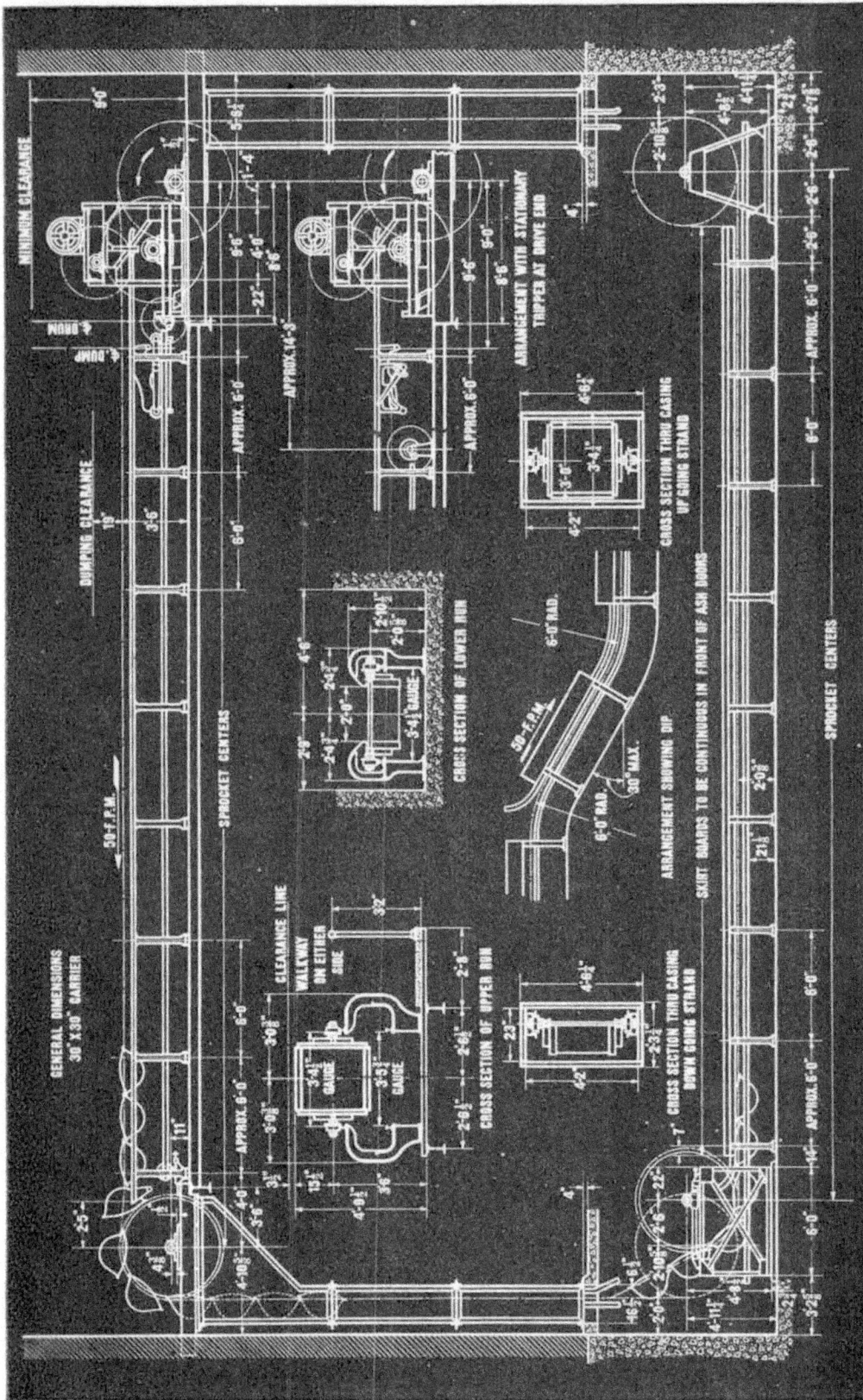

The 30″ x 30″ Carrier handles 125 tons of coal per hour at 50 feet per minute. Walkway cross channels are furnished with upper chairs. In "Cross Section of Lower Run," note protection of Carrier chains by rounded loading skirts. The closest dumping position of Traveling and Stationary Trippers to the Drive Corners is shown above. Plan Views given on next page.

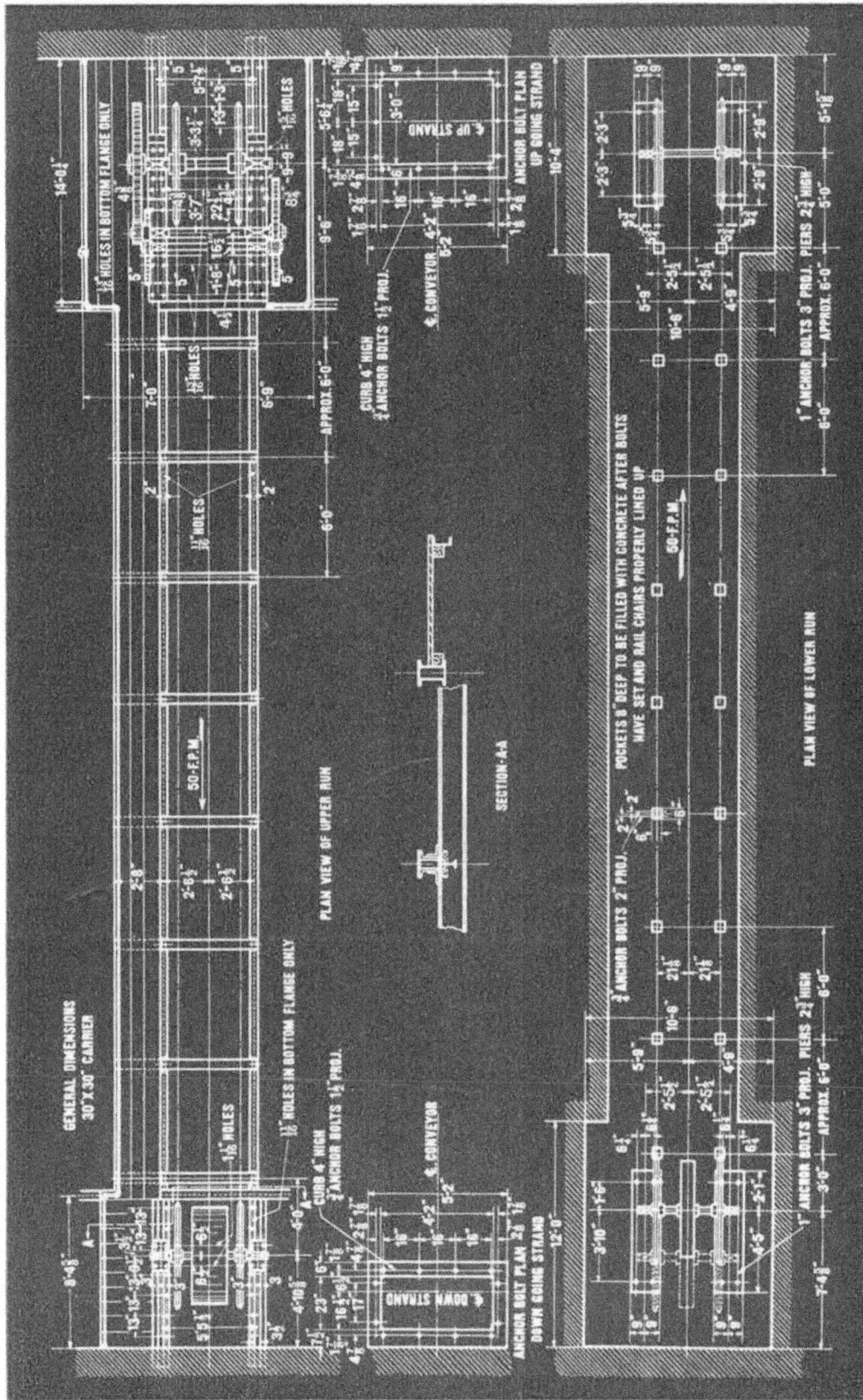

"Plan View of Lower Run" shows the minimum space about the Carrier for proper care and inspection. Where possible increase this clearance along both sides of the Carrier. Anchor bolts in lower corners should be set deep in good cement work. Note the pockets in the Lower Run to be filled with concrete after bolts are in place and rail chairs properly lined up.

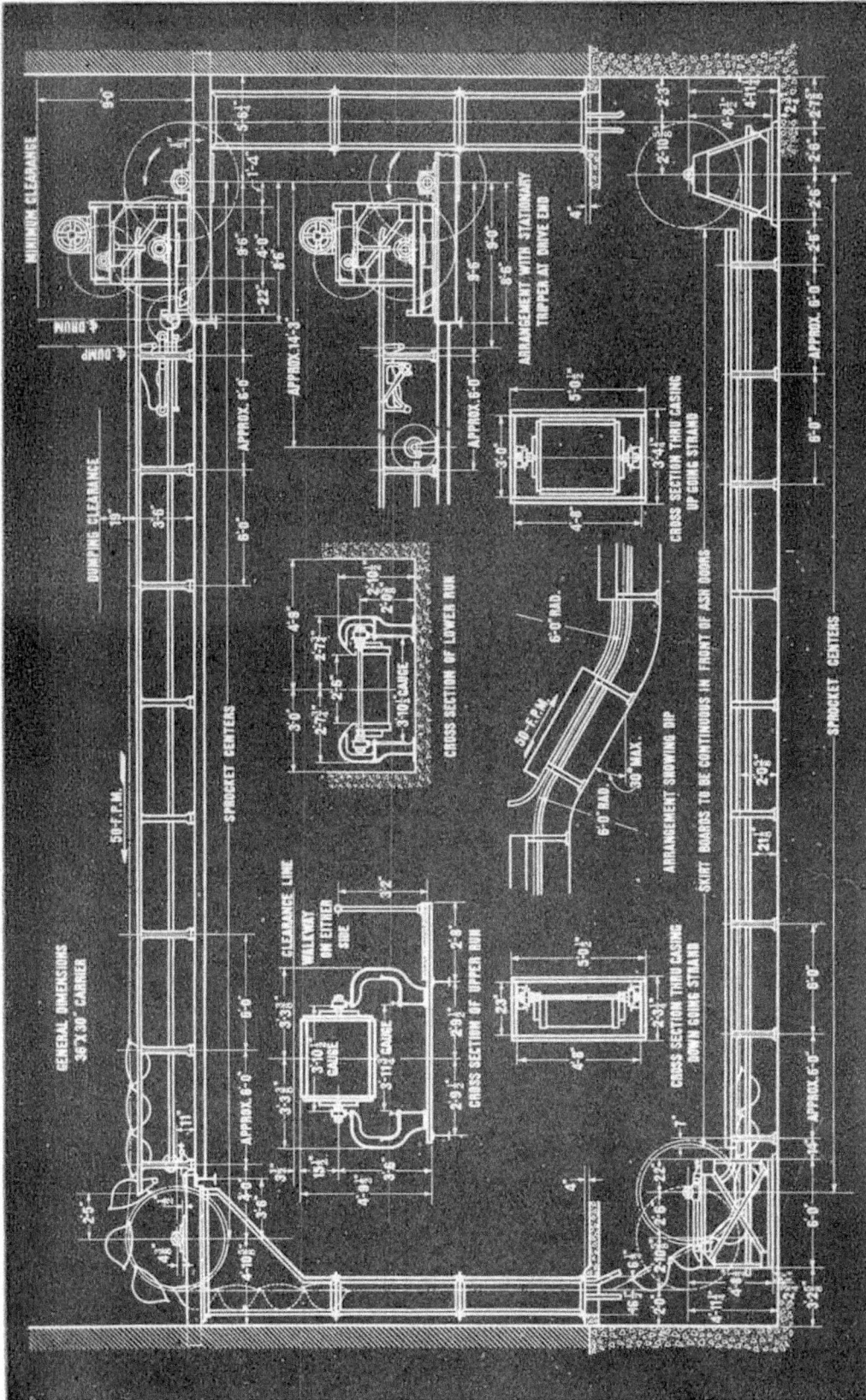

The 36" x 30" Carrier handles 150 tons of Coal per hour at 50 feet per minute.   Walkway cross channels are furnished with upper chairs.   The closest dumping position of Traveling and Stationary Trippers to the Drive Corners is shown above.   Note that pads should be provided under the lower corner frames and casings.   Plan Views given on next page.

"Plan View of Lower Run" shows minimum space about the Carrier for proper care and inspection. Where possible increase this clearance along both sides of the Carrier. Walkway should be provided around the drive end corner to insure proper attention to driving mechanism. Note the Anchor Bolt Plan which gives the location of bolts for fastening the casing to floor.

V-Bucket
Conveyor

Section
2

An Economical Conveyor for Power Plants of moderate capacity—Coaling Stations, Retail Coal Pockets and Yard Storage service.

THE V-Bucket Conveyor so called because of the shape of its buckets, is a combination elevator and scraper conveyor. Primarily, its application is for conditions where material is to be distributed some distance from the vertical lift, or where local conditions will not permit of a vertical elevator being extended a sufficient height to spout the material to the points desired.

An important feature of the V-Bucket Conveyor is its ability to handle large material and is therefore used extensively where the size of the pieces to be handled is beyond the range of the ordinary bucket elevator.

When operating on the horizontal, the V-Bucket Conveyor scrapes or pushes the material along in the trough, discharging by means of valves and chutes. Its use therefore should be confined to handling material of a semi-abrasive nature such as coal, lime, etc., since the wear on the trough and bucket lips when handling material of a more abrasive nature is quite excessive.

Installation views on the following pages show its range of application; its ability to meet various conditions and its working principle as a whole.

The V-Bucket Conveyor as installed in a large Soap Works for supplying coal to bunkers, and which has been in operation for many years. The V-Bucket Conveyor has proven itself both an economical and dependable coal handling system, in this and the many other plants where it has been installed.

Below is shown another Jeffrey V-Bucket Conveyor installed in a boiler house for conveying Coal from railroad cars to overhead bunker storage.

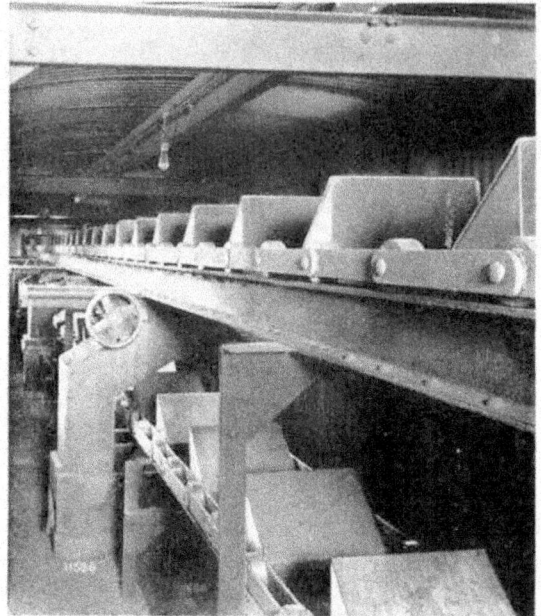

Jeffrey V-Bucket Conveyor serving a large Coal Pocket. The capacity and arrangement of the V-Bucket Conveyor readily lends itself to this class of service, and thus eliminates the necessity of using both an elevator and horizontal conveyor.

The Jeffrey V-Bucket Conveyor also offers an ideal arrangement for the handling of large quantities of coal in outside storage as shown in illustration above. The Conveyor serves both to store and reclaim the coal, which is received by means of a track hopper.

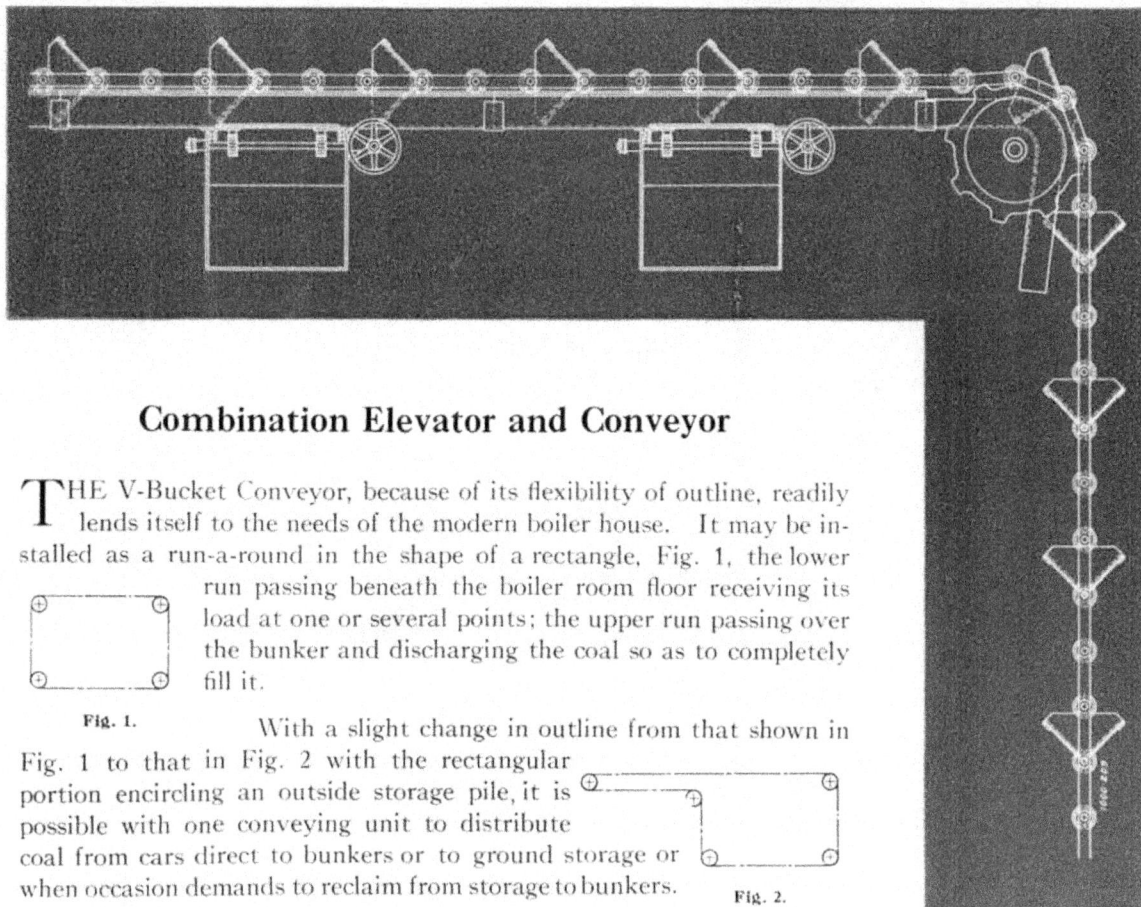

Combination Elevator and Conveyor

THE V-Bucket Conveyor, because of its flexibility of outline, readily lends itself to the needs of the modern boiler house. It may be installed as a run-a-round in the shape of a rectangle, Fig. 1, the lower run passing beneath the boiler room floor receiving its load at one or several points; the upper run passing over the bunker and discharging the coal so as to completely fill it.

Fig. 1.

With a slight change in outline from that shown in Fig. 1 to that in Fig. 2 with the rectangular portion encircling an outside storage pile, it is possible with one conveying unit to distribute coal from cars direct to bunkers or to ground storage or when occasion demands to reclaim from storage to bunkers.

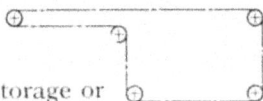

Fig. 2.

Another common V-Bucket installation is that shown in outline Fig. 3 where it is used simply as a vertical elevator discharging its material upon a cross Conveyor where the size of the material handled is larger than could be handled with the ordinary Bucket Elevator.

The larger number of V-Bucket installations however, follow the outline shown in Fig. 4 and for that reason this outline was adopted for the standard conveyors shown on following pages.

Fig. 4.

The vertical lift of forty feet with a horizontal run of eighty feet as given in the standards

Fig. 3.

was found to be a very fair average of hundreds of V-Bucket installations; however, these dimensions may be varied to suit local conditions.

NOTE—For Conveyors beyond the standard limits, specifications and data will be furnished upon request.

Flexibility of the Standard Conveyors

If local conditions are such that an outline as shown in Figs. 1, 2 or 3 would better suit the needs, a standard conveyor may be used and its outline made to conform to the requirements provided the combined vertical lift and horizontal run does not exceed that of the standard. In such an outline as Fig. 2 it is of course necessary to add an additional shaft with its sprockets, bearings etc., of the same sizes as the foot shaft.

41

Operation

The operation of a V-Bucket system, when following an outline such as the standards, is much the same as that of an ordinary Bucket Elevator and Scraper Conveyor. The material is delivered into a boot, is scooped up by the up-going buckets and elevated vertically to the upper corner sprockets where the buckets, being rigidly attached to the chains, turn with the chains thru an angle of 90 degrees and discharge the material into the trough of the upper horizontal run.

To prevent material from spilling as the buckets pass from a vertical to a horizontal position, the trough of the horizontal run is curved around the corner sprockets in such a way that the lips of the buckets just clear the trough bottom in passing around the sprockets. The buckets when operating horizontally act as scrapers and scrape or push the material along in front of them until one of the several openings, controlled by a valve, is reached where it drops thru and is diverted around the return strand of chain and buckets by means of a bifurcated or two way spout.

Capacity and Size of Material.

The capacities of the Standard Conveyors range from 28 to 92 tons of coal per hour when operating at a speed of 100 feet per minute.

The uniform or average size of unsized material varies from dust to pieces 6-inch cubes while the maximum size pieces are twice as large as the average size pieces handled by the various buckets with a limit of 12-inch cubes for the 30" by 24" bucket, but the amount of maximum size pieces should not exceed ten per cent. of the whole.

In the selection of a Standard Conveyor for a given service the first thought is perhaps that of capacity and while it is essential to select a conveyor whose buckets will deliver the amount of material required, the size of the pieces to be handled must not be lost sight of and a bucket that will take care of the maximum size pieces should be chosen irrespective of the capacity requirements. For instance, suppose it is required to handle twenty-five tons of lump coal per hour, some of the lumps as large as 10 or 12 inch cubes. Any one of the standard conveyors would handle this amount of coal nicely but only the largest buckets would handle the large pieces while the capacity of a conveyor with such buckets would be three or four times the requirements. In such instances it is recommended that the speed of the conveyor be reduced proportionately, thereby materially increasing the life of the equipment.

Power Requirements.

The amount of power required to operate the various conveyors is given in the specification tables for each conveyor and is listed under the second countershaft. The value given in each instance is for that particular conveyor with a vertical lift of 40 feet and a horizontal run of 80 feet operating at a speed of 100 feet per minute and handling the capacity listed. It is the power required at the second countershaft which has a keyseated extension to receive purchasers pulley or cut tooth gear if it is desired to direct-connect to a motor. To find size of motor to use add five percent to the value given in the tables.

While the horsepower listed varies with the capacity and centers of the conveyor, it is not recommended that the motor horsepower be decreased as these values decrease. However, as the horsepower is a direct function of the speed at which the conveyor is operating, it may be decreased proportionately with any decrease in speed.

Kind and Location of Valves.

The type of valves used with Jeffrey Standard V-Bucket Conveyors, unless otherwise specified by the purchaser, are the bevel gear operated rack and pinion type as shown on page 96. The operating hand wheel may be placed in a horizontal position as in Fig. 3, or vertical as in Fig. 4. The construction of the valve guides is such as to eliminate the possibility of fine material lodging on same and causing the valve plate to bind or stick.

In locating valves in the upper horizontal trough they should be so spaced as to serve the bunker or storage pile to the best advantage. For instance, in the case of conveyor operating over a bunker in a boiler house they should be spaced sufficiently close together to satisfactorily fill the bunker and at the same time be over a stoker spout, thereby insuring a direct supply of coal to the stokers.

Index to Standard V-Bucket Conveyors

Conveyor No.	Average Size Mat'l to be handled Inches	Maximum size pieces Inches	Capacity in Tons per Hour Coal	Size of Bucket—In.			Chain	Page Number
				Length	Width	Depth		
3249	3½	7	28	18	14	7	526 Vul	44
3250	4	8	42	20	16	8	526 Vul	44
3251	3½	7	28	18	14	7	516 F & R	46
3252	4	8	42	20	16	8	516 F & R	46
3253	3½	7	28	18	14	7	126C MR	48
3254	4	8	42	20	16	8	126C MR	48
3255	3½	7	28	18	14	7	951 STR	50
3256	4	8	42	20	16	8	951 STR	50
3257	4	8	31	20	16	8	558 Vul	52
3258	4½	9	47	24	18	9	558 Vul	52
3259	5	10	63	26	20	10	558 Vul	52
3260	4	8	31	20	16	8	518 F & R	54
3261	4½	9	47	24	18	9	518 F & R	54
3262	5	10	63	26	20	10	518 F & R	54
3263	4	8	28	20	16	8	276 STR	56
3264	4½	9	41	24	18	9	276 STR	56
3265	5	10	56	26	20	10	276 STR	56
3266	5	10	56	26	20	10	180 STR	58
3267	6	12	92	30	24	12	180 STR	58
3268	6	12	92	30	24	12	182½ STR	60

How to Select Conveyors from Tables

THE Index and Table of Capacities given above together with tables of specifications on following pages make it a simple matter to select the Conveyor to suit requirements. While the data given is based on the installation shown in Figure 4 on page 41, the same can be used for installations following other outlines.

In selecting a Conveyor, the "Average Size Material to be handled", "Maximum Size Pieces", "Capacity in Tons per Hour" and the "Lift and Run of Conveyor", must first be determined. For example, let us assume the Average Size of Material is 4-inch pieces, the Maximum Size 10-inch pieces, the Capacity Requirement 40 tons per hour, the Lift (center to center of sprockets) is 35 feet and the Run (center to center of sprockets) is 86 feet. In the second column of the table above it will be noted that there are 7 sizes that permit of an average size piece of 4 inches, but only four sizes are listed for a maximum piece of 10 inches.

In the capacity column it will be found that the majority of sizes will handle 40 tons per hour, however, only those having a 26" x 20" x 10" bucket will handle 10 inch lumps. Therefore either Conveyors No. 3259, 3262, 3265 or 3266 will meet the requirements providing the centers do not exceed those of the standard conveyors. The 86 feet of run is in excess of that listed in the table by 6 feet, but since the lift is less by 5 feet, any one of the four Conveyors may be used, because a decrease of one foot in the Lift allows an addition of 2 feet on the Run. Referring to pages indicated opposite the Conveyors noted you will find Specifications and Dimensions and by comparison of same it is an easy matter to select the Conveyor best suited.

Using No. 526 Vulcan Chain

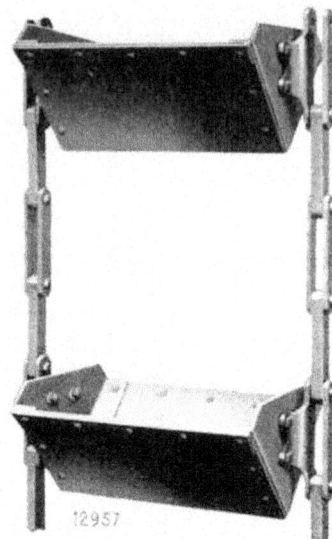

Installation View of Jeffrey V-Bucket Conveyor using No. 526 Vulcan Chain.

Specifications

MATERIAL: Coal or similar material weighing approximately 50 lbs. per cu. ft.
SPEED: 100 Feet per minute. MAX. CENTERS: 40 ft. vertical, 80 ft. horizontal

Number of Elevator	3249	3250	Number of Elevator	3249	3250
Size of Material—Inches			**1st Countershaft—In.—Cont'd**		
Uniform or Average of un-sized Material	3½	4	Diameter of Shaft	2⁷⁄₁₆	2¹⁵⁄₁₆
			Revolutions per Minute	56	56
Maximum size not to exceed			Diameter of Gear	25.09	25.09
10% of whole	7	8	Pitch of Gear	1¼	1¼
			Face of Gear	3	3
Capacity—Tons per Hour	28	42			
			2nd Countershaft—Inches		
Chain			Diameter of Pinion	6.01	6.01
Number	526 Vul.	526 Vul.	Face of Pinion	3¼	3¼
Attachments	V E-1	V E-1	Diameter of Shaft	1¹⁵⁄₁₆	1¹⁵⁄₁₆
Pitch—Inches	6	6	Revolutions per Minute	235	235
Working Strength—Lbs.	1640	1640	H. P. Required—Max. Ctrs.	7.5	9.5
Buckets—Inches			**Corner Shaft—Inches**		
Length	18	20	Diameter Shaft—upper corner	2⁷⁄₁₆	2⁷⁄₁₆
Width	14	16	Diameter Sprocket—upper		
Depth	7	8	corner	23⅛	27
Gauge	10	10	Diameter Shaft—lower corner	1¹⁵⁄₁₆	1¹⁵⁄₁₆
Spacing	24	24	Diam. Sprocket—lower corner	19⅜	23⅜
Headshaft—Inches			**Foot Shaft—Inches**		
Diameter of Shaft	2¹⁵⁄₁₆	3⁷⁄₁₆	Diameter of Shaft	1¹⁵⁄₁₆	1¹⁵⁄₁₆
Revolutions per Minute	11.1	11.1	Diameter of Sprocket	19⅜	23⅜
Diameter of Sprocket	34⅝	34⅝			
Diameter of Gear	35.82	35.82	**Approx. Shipping Wgt.—Lbs.**		
Pitch of Gear	1½	1½	Chain and Buckets per Foot		
Face of Gear	4	4	Centers	51	57
			Machinery Terminals	2260	2630
1st Countershaft—Inches			Casing Terminals	4000	4580
Diameter of Pinion	7.22	7.22	Casing per Foot	127	137
Face of Pinion	4½	4½	Trough per Foot	32	35

Using No. 526 Vulcan Chain

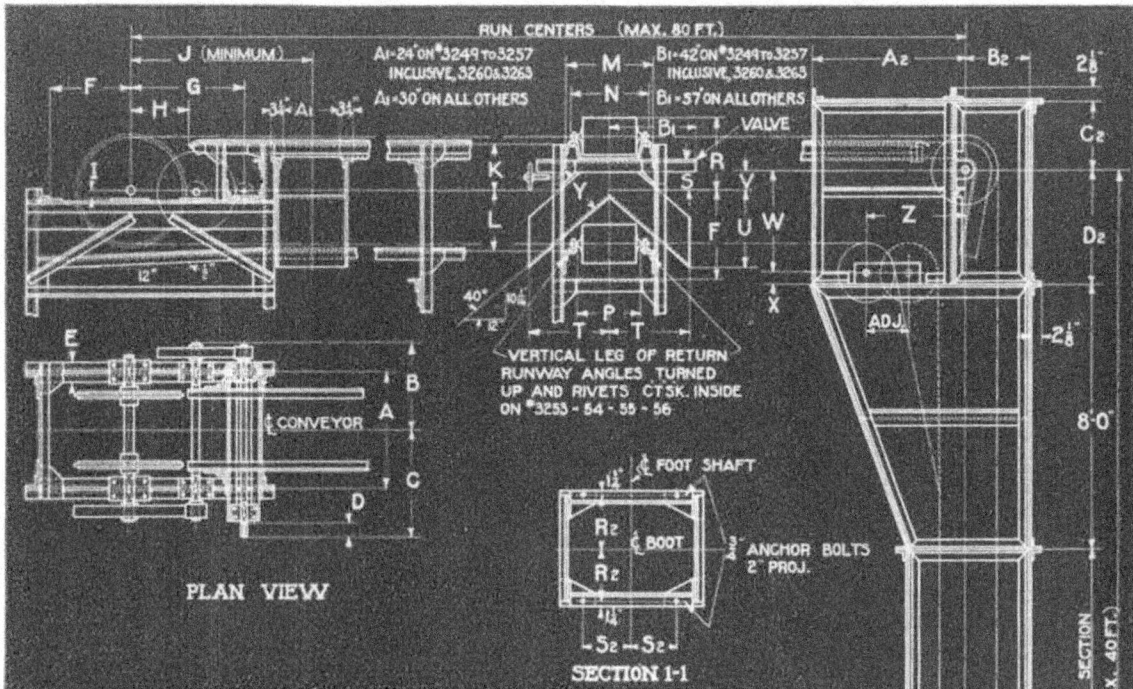

PLAN VIEW

SECTION 1-1

General Dimensions

Dimensions	ELEVATOR No. 3249	ELEVATOR No. 3250	Dimensions	ELEVATOR No. 3249	ELEVATOR No. 3250
A	36¾	40½	W	33¼	33¼
B	27⅛	30	X	2¾	2¾
C	36	39	Y	11½	12
D	6	6	Z	28⅞	30⅞
E	6¼	6¼	A2	49⅜	51⅜
F	28¾	29¾	B2	22⅝	26⅝
G	37⅛	37⅛	C2	21	24
H	18	18	D2	36	36
I	3⅝	4	E2	10'-0"	10'-0"
J	60	60	F2	29¾	31¾
K	15¾	15¾	G2	40	48
L	20	20	H2	1½	1½
M	27	29	J2	10'-0"	10'-0"
N	24¾	26¾	K2	7½	10½
P	20¾	22¾	L2	12	12
R	24⅛	25⅛	M2	20	24
S	8¼	7¼	N2	19	23
T	25	25	P2	21⅛	25⅛
U	18½	19½	R2	16¾	17¾
V	5¾	3⅞	S2	15	19

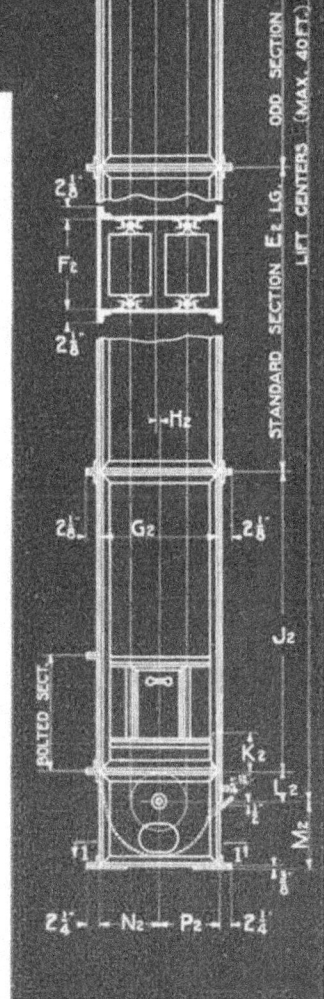

V-Bucket Conveyors — JEFFREY=
Using No 516 F. and R. Chain

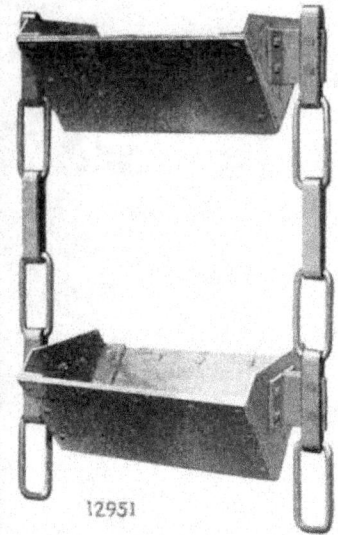

Installation View of Jeffrey V-Bucket Conveyor using No. 516 Flat and Round Chain.

Specifications

MATERIAL: Coal or similar material weighing approximately 50 lbs. per cu. ft.
SPEED: 100 Feet per minute. MAX. CENTERS: 40 ft. vertical, 80 ft. horizontal.

Number of Elevator	3251	3252	Number of Elevator	3251	3252
Size of Material—Inches			**1st Countershaft—In.—Cont'd**		
Uniform or Average of un-sized Material	3½	4	Diameter of Shaft	$2\frac{7}{16}$	$2\frac{13}{16}$
			Revolutions per Minute	56	56
Maximum size not to exceed 10% of whole	7	8	Diameter of Gear	25.09	25.09
			Pitch of Gear	1¼	1¼
Capacity—Tons per Hour	28	42	Face of Gear	3	3
Chain			**2nd Countershaft—Inches**		
Number	516 F & R	516 F & R	Diameter of Pinion	6.01	6.01
Attachments	V E-1	V E-1	Face of Pinion	3¼	3¼
Pitch—Inches	6	6	Diameter of Shaft	$1\frac{13}{16}$	$1\frac{13}{16}$
Working Strength—Lbs.	3400	3400	Revolutions per Minute	235	235
			H. P. Required—Max. Ctrs.	7.0	9.0
Buckets—Inches			**Corner Shafts—Inches**		
Length	18	20	Diameter Shaft—upper corner	$2\frac{7}{16}$	$2\frac{7}{16}$
Width	14	16	Diameter Sprocket—upper corner	23⅜	27⅛
Depth	7	8	Diameter Shaft—lower corner	$1\frac{13}{16}$	$1\frac{13}{16}$
Gauge	10	10	Diam. Sprocket—lower corner	19⅝	23⅜
Spacing	24	24			
Headshaft—Inches			**Foot Shaft—Inches**		
Diameter of Shaft	$2\frac{13}{16}$	$3\frac{7}{16}$	Diameter of Shaft	$1\frac{13}{16}$	$1\frac{13}{16}$
Revolutions per Minute	11.1	11.1	Diameter of Sprocket	19⅝	23⅜
Diameter of Sprocket	34¾	34¾			
Diameter of Gear	35.82	35.82	**Approx. Shipping Wgt.—Lbs.**		
Pitch of Gear	1½	1½	Chain and Buckets per Ft. Centers	47	53
Face of Gear	4	4	Machinery Terminals	2210	2580
1st Countershaft—Inches			Casing Terminals	4000	4580
Diameter of Pinion	7.22	7.22	Casing per Foot	127	137
Face of Pinion	4½	4½	Trough per Foot	32	35

Using No. 516 F. and R. Chain

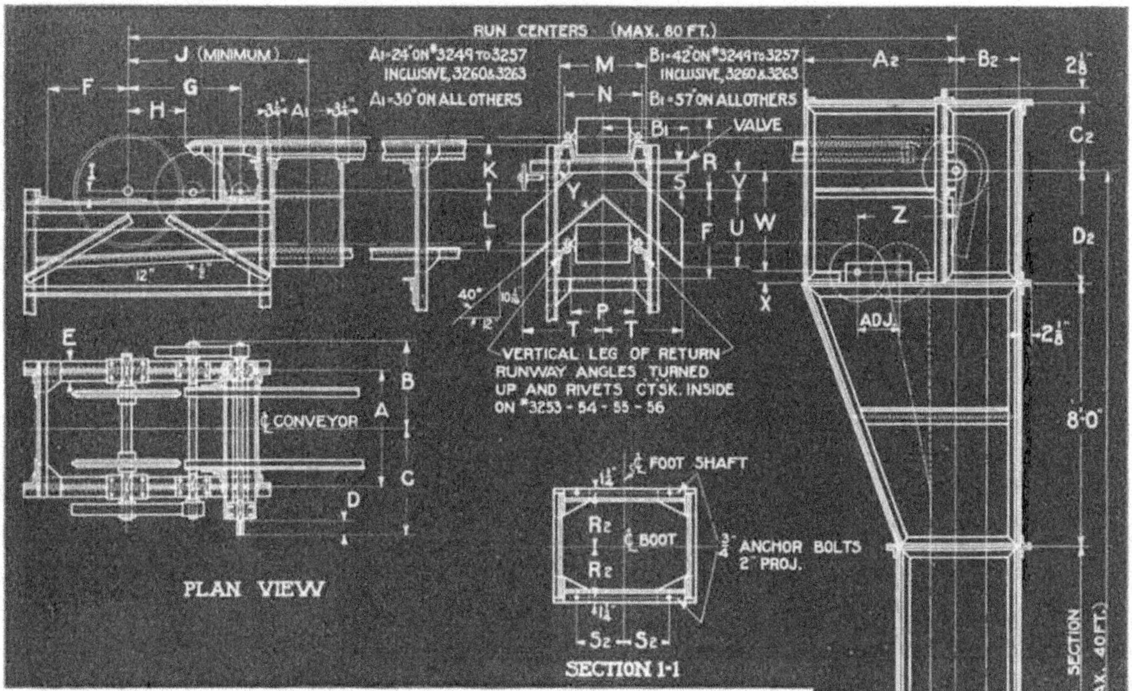
RUN CENTERS (MAX. 80 FT.)
J (MINIMUM)
A1-24" ON #3249 TO 3257 INCLUSIVE, 3260 & 3263
A1-30" ON ALL OTHERS
B1-42" ON #3249 TO 3257 INCLUSIVE, 3260 & 3263
B1-57" ON ALL OTHERS
VALVE
VERTICAL LEG OF RETURN RUNWAY ANGLES TURNED UP AND RIVETS CTSK. INSIDE ON #3253 - 54 - 55 - 56
℄ CONVEYOR
℄ FOOT SHAFT
℄ BOOT
¾" ANCHOR BOLTS 2" PROJ.
PLAN VIEW
SECTION 1-1
ADJ.

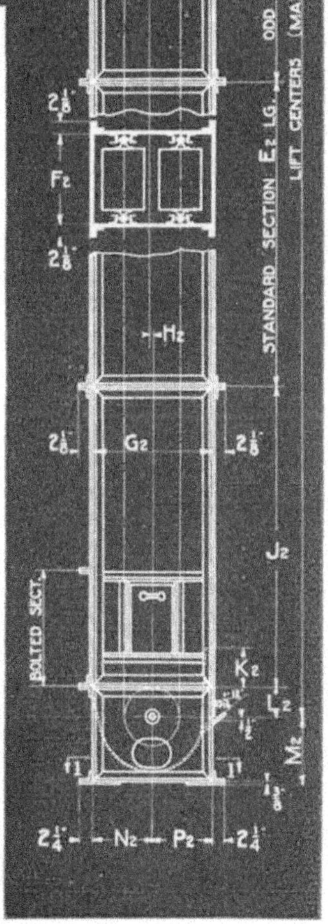
ODD SECTION
LIFT CENTERS (MAX. 40 FT.)
STANDARD SECTION E2 LG.
BOLTED SECT.

General Dimensions

Dimensions	ELEVATOR No. 3251	ELEVATOR No. 3252	Dimensions	ELEVATOR No. 3251	ELEVATOR No. 3252
A	36¾	40½	W	33¼	33¼
B	27⅛	30	X	2¾	2¾
C	36	39	Y	11½	12
D	6	6	Z	28⅞	30⅞
E	6¼	6¼	A2	49⅜	51⅜
F	28¾	29¾	B2	22⅝	26⅝
G	37⅛	37⅛	C2	21	24
H	18	18	D2	36	36
I	3⅝	4	E2	10′-0″	10′-0″
J	60	60	F2	29¾	31¾
K	15¾	15¾	G2	40	48
L	20	20	H2	1½	1½
M	28	30	J2	10′-0″	10′-0″
N	25¾	27¾	K2	7½	10½
P	20¾	22¾	L2	12	12
R	24⅛	25⅛	M2	20	24
S	8¼	7¼	N2	19	23
T	25	25	P2	21⅛	25⅛
U	18½	19½	R2	16¾	17¾
V	5¾	3⅞	S2	15	19

Using No. 126-C M. R. Chain

Installation View of Jeffrey V-Bucket Conveyor using No. 126C Malleable Roller Chain.

Specifications

MATERIAL: Coal or similar material weighing approximately 50 lbs. per cu. ft.

SPEED: 100 Feet per minute. MAX. CENTERS: 40 ft. vertical, 80 ft. horizontal.

Number of Elevator	3253	3254	Number of Elevator	3253	3254
Size of Material—Inches			**1st Countershaft—In.–Cont'd**		
Uniform or Average of un-sized Material	3½	4	Revolutions per Minute	56	56
			Diameter of Gear	25.09	25.09
Maximum size not to exceed 10% of whole	7	8	Pitch of Gear	1¼	1¼
			Face of Gear	3	3
Capacity—Tons per Hour	28	42	**2nd Countershaft—Inches**		
			Diameter of Pinion	6.01	6.01
Chain			Face of Pinion	3¼	3¼
Number	126 C M R	126 C M R	Diameter of Shaft	1¹⁵⁄₁₆	1¹⁵⁄₁₆
Attachments	V E-1	V E-1	Revolutions per Minute	235	235
Pitch—Inches	6	6	H. P. Required—Max. Ctrs.	7.0	9.0
Working Strength—Lbs.	3100	3100			
			Corner Shaft—Inches		
Buckets—Inches			Diameter Shaft—upper corner	2⁷⁄₁₆	2⁷⁄₁₆
Length	18	20	Diameter Sprocket—upper corner	23¼	27
Width	14	16	Diameter Shaft—lower corner	1¹⁵⁄₁₆	1¹⁵⁄₁₆
Depth	7	8	Diameter Sprocket—lower corner	19⅜	23¼
Gauge	10	10			
Spacing	24	24	**Foot Shaft—Inches**		
			Diameter of Shaft	1¹⁵⁄₁₆	1¹⁵⁄₁₆
Headshaft—Inches			Diameter of Sprocket	19⅜	23¼
Diameter of Shaft	2¹⁵⁄₁₆	3⁷⁄₁₆			
Revolutions per Minute	11.1	11.1	**Approx. Shipping Wgt.—Lbs.**		
Diameter of Sprocket	34⅝	34⅝	Chain and Buckets per Ft. Centers	59	65
Diameter of Gear	35.82	35.82	Machinery Terminals	2430	2865
Pitch of Gear	1½	1½	Casing Terminals	4000	4580
Face of Gear	4	4	Casing per Foot	127	137
			Trough per Foot	32	35
1st Countershaft—Inches					
Diameter of Pinion	7.22	7.22			
Face of Pinion	4½	4½			
Diameter of Shaft	2⁷⁄₁₆	2¹⁵⁄₁₆			

Using No. 126-C M. R. Chain

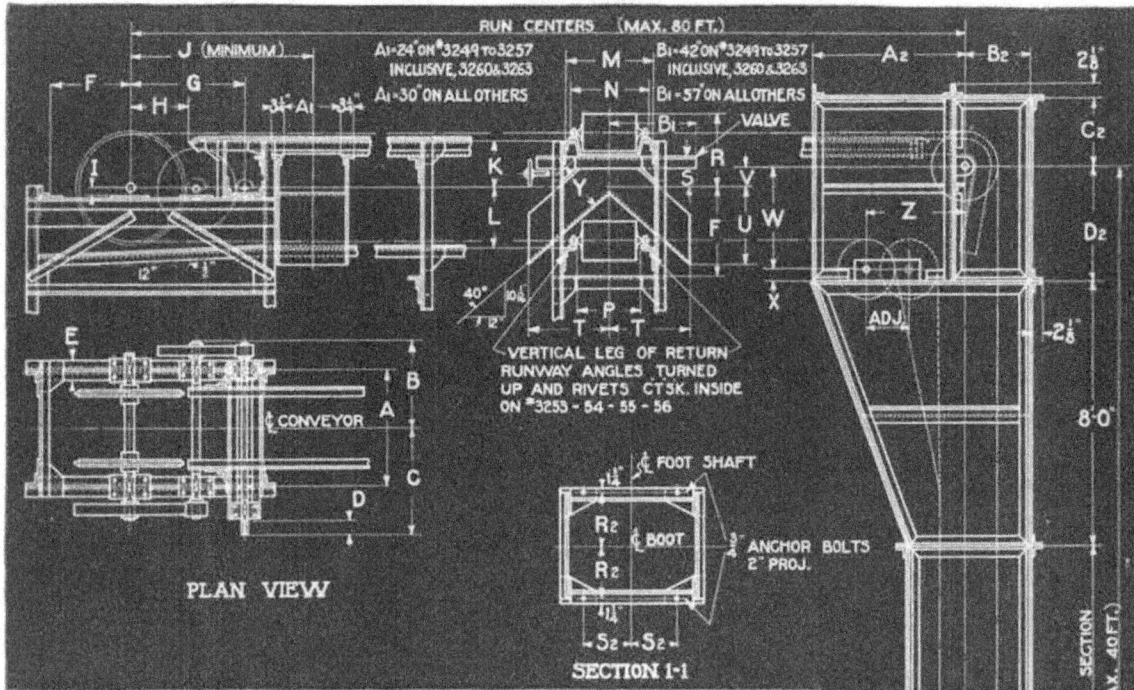

PLAN VIEW

SECTION 1-1

General Dimensions

Dimen- sions	ELEVATOR No.		Dimen- sions	ELEVATOR No.	
	3253	3254		3253	3254
A	37½	41¼	W	33¼	33¼
B	27½	30⅜	X	2¾	2¾
C	37	39	Y	11½	12
D	6	6	Z	28⅞	30⅞
E	6¼	6¼	A2	49⅜	51⅜
F	28¾	29¾	B2	22⅝	26⅝
G	37⅛	37⅛	C2	21	24
H	18	18	D2	36	36
I	3⅝	4	E2	10'-0"	10'-0"
J	60	60	F2	29¾	31¾
K	15	15	G2	40	48
L	20¾	20¾	H2	1½	1½
M	29	31	J2	10'-0"	10'-0"
N	26¾	28¾	K2	7½	10½
P	21½	23½	L2	12	12
R	24⅛	25⅛	M2	20	24
S	8¼	7¼	N2	19	23
T	25	25	P2	21⅛	25⅛
U	18½	19½	R2	16¾	17¾
V	5¾	3⅞	S2	15	19

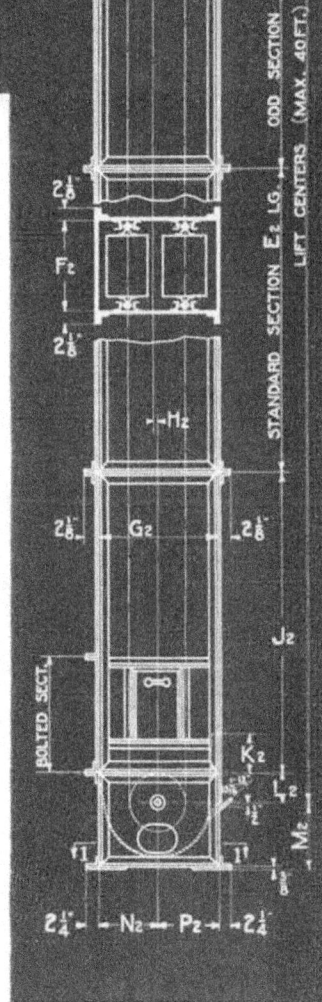

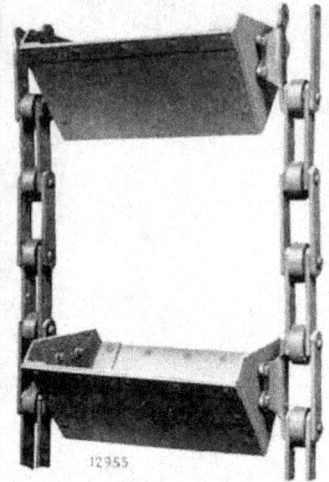

Installation View of Jeffrey V-Bucket Conveyor using No. 951 Steel Thimble Roller Chain.

Specifications

MATERIAL: Coal or similar material weighing approximately 50 lbs. per cu. ft.

SPEED: 100 Feet per minute. MAX. CENTERS: 40 ft. vertical, 80 ft. horizontal.

Number of Elevator	3255	3256	Number of Elevator	3255	3256
Size of Material—Inches			**1st Countershaft—In.—Cont'd**		
Uniform or Average of un-sized Material	3½	4	Revolutions per Minute	56	56
			Diameter of Gear	25.09	25.09
Maximum size not to exceed			Pitch of Gear	1¼	1¼
10% of whole	7	8	Face of Gear	3	3
Capacity—Tons per Hour	28	42	**2nd Countershaft—Inches**		
			Diameter of Pinion	6.01	6.01
Chain			Face of Pinion	3¼	3¼
Number	951 S T R	951 S T R	Diameter of Shaft	1¹³⁄₁₆	1¹³⁄₁₆
Attachments	V E-1	V E-1	Revolutions per Minute	235	235
Pitch—Inches	6	6	H. P. Required—Max. Ctrs.	6.5	8.5
Working Strength—Lbs.	3750	3750	**Corner Shafts—Inches**		
			Diameter Shaft—upper corner	2⁷⁄₁₆	2⁷⁄₁₆
Buckets—Inches			Diameter Sprocket—upper		
Length	18	20	corner	23¼	27
Width	14	16	Diameter Shaft—lower corner	1¹⁵⁄₁₆	1¹⁵⁄₁₆
Depth	7	8	Diameter Sprocket—lower		
Gauge	10	10	corner	19⅜	23¼
Spacing	24	24	**Foot Shaft—Inches**		
Headshaft—Inches			Diameter of Shaft	1¹⁵⁄₁₆	1¹⁵⁄₁₆
Diameter of Shaft	2¹⁵⁄₁₆	3⁷⁄₁₆	Diameter of Sprocket	19⅜	23¼
Revolutions per Minute	11.1	11.1			
Diameter of Sprocket	34⅝	34⅝	**Approx. Shipping Wgt.—Lbs.**		
Diameter of Gear	35.82	35.82	Chain and Buckets per Ft.		
Pitch of Gear	1½	1½	Centers	67	73
Face of Gear	4	4	Machinery Terminals	2450	2890
1st Countershaft—Inches			Casing Terminals	4000	4580
Diameter of Pinion	7.22	7.22	Casing per Foot	127	137
Face of Pinion	4½	4½	Trough per Foot	32	35
Diameter of Shaft	2⁷⁄₁₆	2¹⁵⁄₁₆			

V-Bucket Conveyors
Using No. 951 S. T. R. Chain

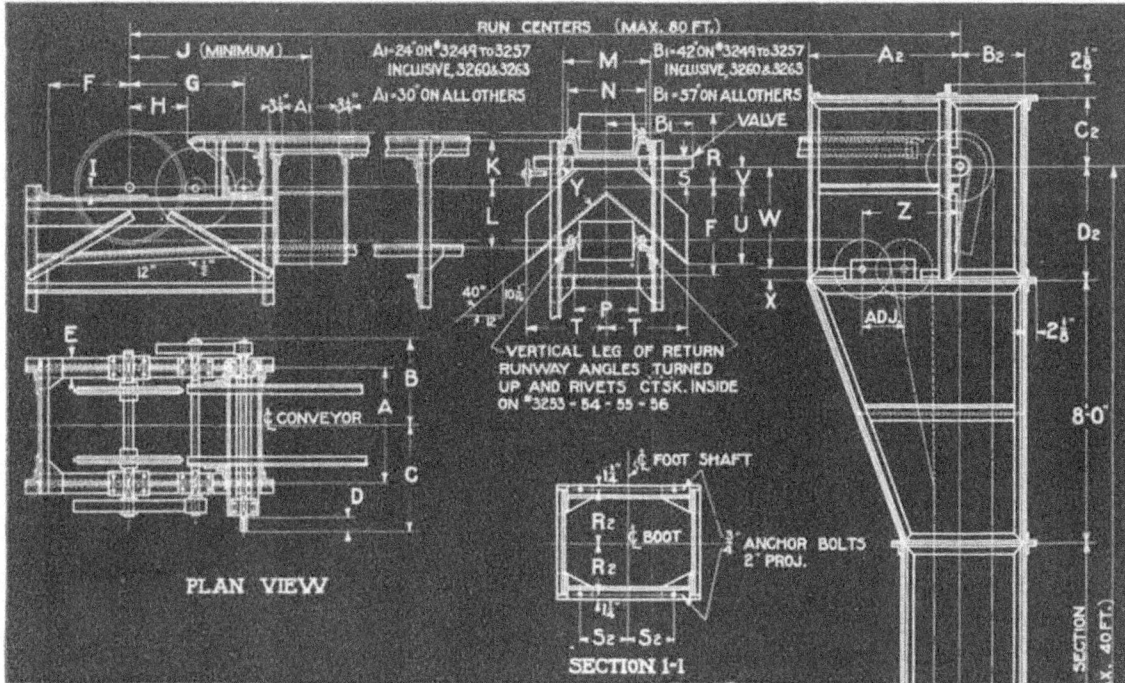

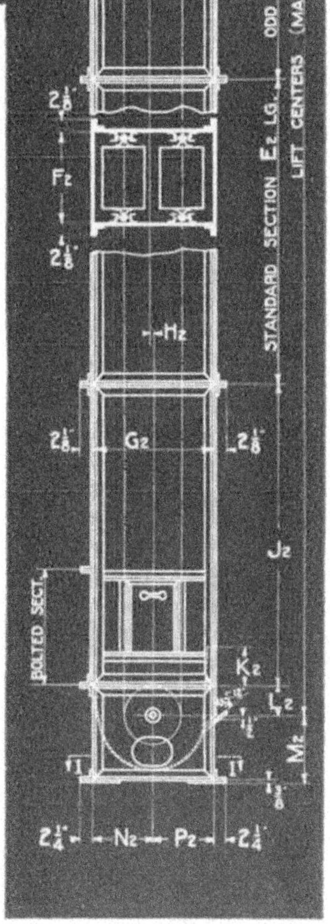

PLAN VIEW

SECTION I-I

General Dimensions

Dimen-sions	ELEVATOR No.		Dimen-sions	ELEVATOR No.	
	3255	3256		3255	3256
A	37½	41¼	W	33¼	33¼
B	27½	30⅜	X	2¾	2¾
C	37	39	Y	11½	12
D	6	6	Z	28⅞	30⅞
E	6¼	6¼	A2	49⅜	51⅜
F	28¾	29¾	B2	22⅝	26⅝
G	37⅛	37⅛	C2	21	24
H	18	18	D2	36	36
I	3⅝	4	E2	10'-0"	10'-0"
J	60	60	F2	29¾	31¾
K	15	15	G2	40	48
L	20¾	20¾	H2	1½	1½
M	29	31	J2	10'-0"	10'-0"
N	26¾	28¾	K2	7½	10½
P	21½	23½	L2	12	12
R	24⅛	25⅛	M2	20	24
S	8¼	7¼	N2	19	23
T	25	25	P2	21⅛	25⅛
U	18½	19½	R2	16¾	17¾
V	5¾	3⅞	S2	15	19

V-Bucket Conveyors ═══ ═JEFFREY═
Using No. 558 Vulcan Chain

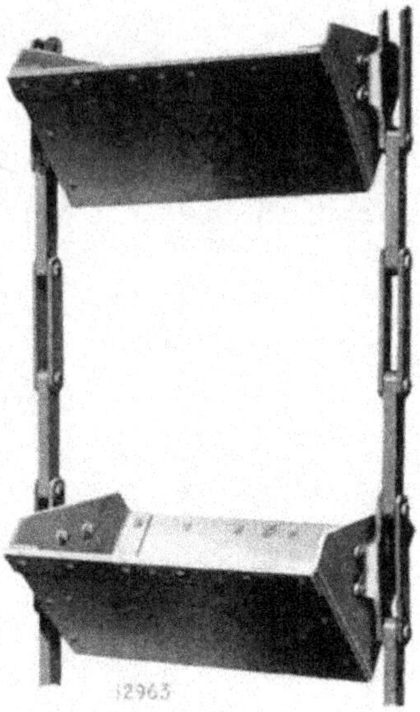

Installation View of Jeffrey V-Bucket Conveyor using No. 558 Vulcan Chain.

Specifications

MATERIAL: Coal or similar material weighing approximately 50 lbs. per cu. ft.
SPEED: 100 Feet per minute. MAX. CENTERS: 40 ft. vertical, 80 ft. horizontal.

No. of Elevator	3257	3258	3259	No. of Elevator	3257	3258	3259
Size of Material—In.				**1st Countershaft— Inches—Cont'd**			
Uniform or Avg. of unsized material ..	4	4½	5	Diam. of Gear..........	25.09	32.00	32.00
Max. size not to exceed 10% of whole..	8	9	10	Pitch of Gear...........	1¼	1½	1½
				Face of Gear	3	4	4
Capacity—Tons per Hour	31	47	63	**2nd Countershaft— Inches**			
				Diam. of Pinion.......	6.01	7.22	7.22
Chain				Face of Pinion........	3¼	4½	4½
Number..............	558 Vul.	558 Vul.	558 Vul.	Diam. of Shaft........	1¹⁵⁄₁₆	2⁷⁄₁₆	2⁷⁄₁₆
Attachments	V E-1	V E-1	V E-1	Rev. per minute.......	225	230	230
Pitch—In.............	8	8	8	H. P. Required—			
Working Strength— Lbs.	2250	2250	2250	Max. Centers........	8.5	11.5	14.0
Buckets—In.				**Corner Shafts—In.**			
Length.............	20	24	26	Diam. Shaft, upper corner...............	2⁷⁄₁₆	2¹⁵⁄₁₆	2¹⁵⁄₁₆
Width..............	16	18	20	Diam. Sprocket— upper corner..........	25⅞	30⅞	30⅞
Depth..............	8	9	10	Diam. Shaft—lower corner	1¹⁵⁄₁₆	2⁷⁄₁₆	2⁷⁄₁₆
Gauge.............	10	3⁄₁₆	3⁄₁₆	Diam. Sprocket— lower corner..........	25⅞	25⅞	30⅞
Spacing...........	32	32	32				
Headshaft—In.				**Foot Shaft—In.**			
Diam. of Shaft......	3⁷⁄₁₆	3¹⁵⁄₁₆	3¹⁵⁄₁₆	Diam. of Shaft.........	1¹⁵⁄₁₆	2⁷⁄₁₆	2⁷⁄₁₆
Rev. per minute.......	10.7	8.3	8.3	Diam. of Sprocket....	25⅞	25⅞	30⅞
Diam. of Sprocket...	36	46⅝	46⅝				
Diam. of Gear......	35.82	48.47	48.47	**Approx. Shipping Weight—Lbs.**			
Pitch of Gear...........	1½	1¾	1¾	Chain and Buckets per Ft. Ctrs.......	56	76	84
Face of Gear	4	5½	5½	Mach. Terminals......	2920	4240	4464
1st Countershaft—In.				Casing Terminals ...	4580	5200	5775
Diam. of Pinion......	7.22	7.86	7.86	Casing per Foot......	143	151	153
Face of Pinion........	4½	6	6	Trough per Foot	35	39	42
Diam. of Shaft........	2¹⁵⁄₁₆	2¹⁵⁄₁₆	2¹⁵⁄₁₆				
Rev. per minute.......	54	52	52				

Using No. 558 Vulcan Chain

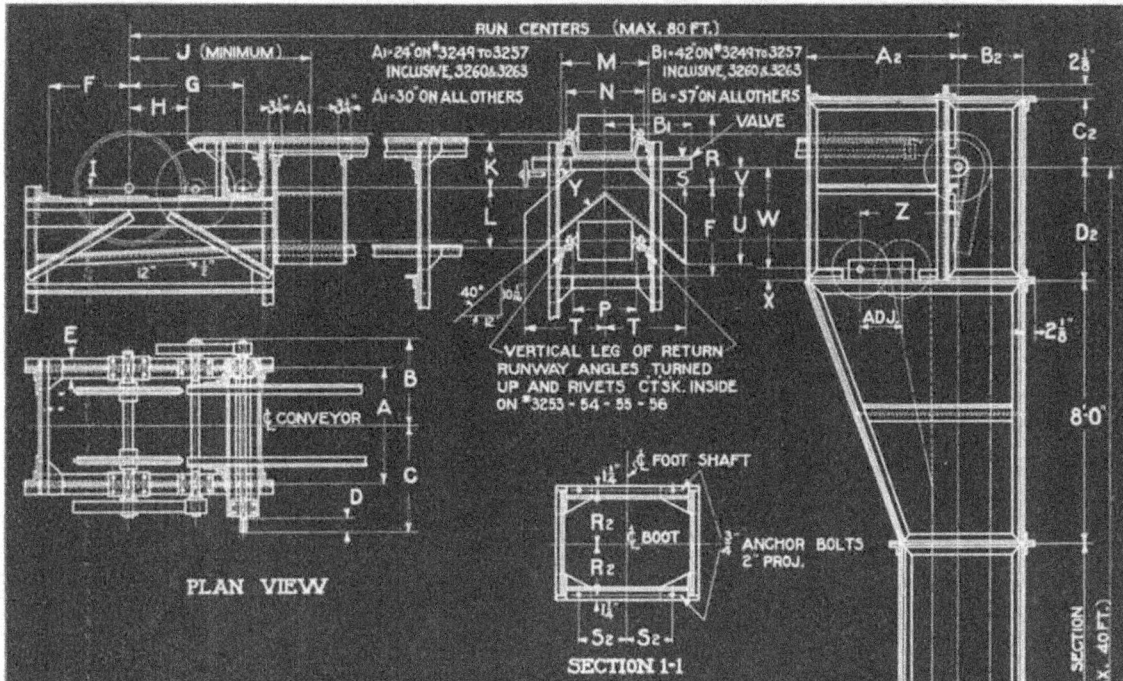

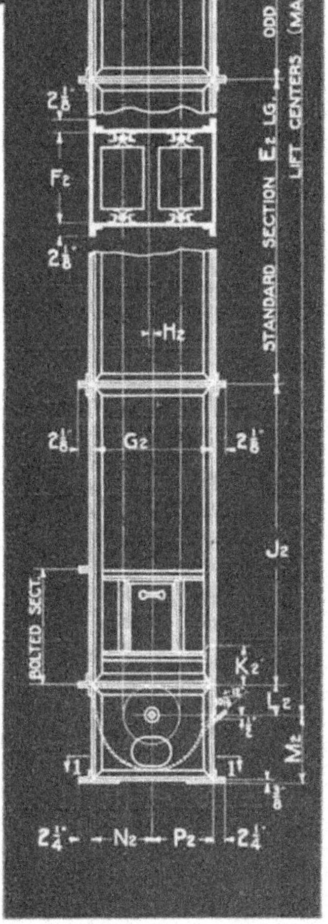

PLAN VIEW

SECTION 1-1

General Dimensions

Dimen-sions	ELEVATOR No.			Dimen-sions	ELEVATOR No.		
	3257	3258	3259		3257	3258	3259
A	40¾	47	49	W	35¼	44⅜	46⅜
B	30⅛	33¼	34¼	X	2¾	3⅛	3⅛
C	39	44	45	Y	12	18	18½
D	6	6	6	Z	37⅜	41⅞	45⅞
E	6¼	8¼	8¼	A2	63⅜	69⅞	73⅞
F	30½	37¼	38¼	B2	26⅝	30⅛	31⅛
G	37⅛	47¾	47¾	C2	24	26½	28½
H	19	22	22	D2	38	47½	49½
I	4	4⅝	4⅝	E2	8'-0"	8'-0"	10'-0"
J	60	75	75	F2	31¾	35¾	37¾
K	16⅛	21¼	21¼	G2	50	52	60
L	20½	26⅜	26⅜	H2	0	2⅜	0
M	29½	33½	35½	J2	7'-8"	7'-8"	9'-4"
N	27¼	31¼	33¼	K2	8½	8½	9½
P	23	27	29	L2	14	16½	18½
R	25⅝	31⅞	32⅞	M2	24	25½	29½
S	7½	11¾	10¾	N2	23½	24⅜	29
T	25	30	30	P2	26⅝	27¾	31⅛
U	19⅛	25⅛	26⅛	R2	17¾	19¾	20¾
V	5⅛	7⅝	7⅝	S2	19	20	25

V-Bucket Conveyors
Using No. 518 F. and R. Chain

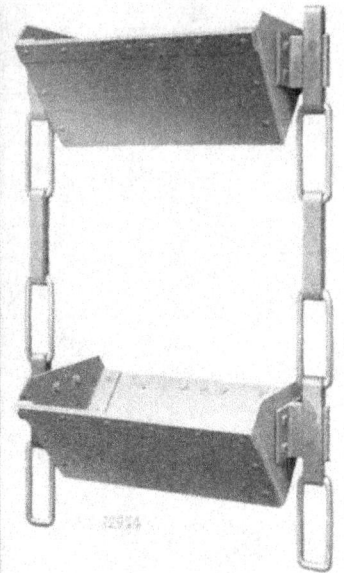

Installation View of Jeffrey V- Bucket Conveyor using No. 518 Flat and Round Chain.

Specifications

MATERIAL: Coal or similar material weighing approximately 50 lbs. per cu. ft.
SPEED: 100 Feet per minute. MAX. CENTERS: 40 ft. vertical, 80 ft. horizontal.

No. of Elevator	3260	3261	3262
Size of Material—In.			
Uniform or Avg. of unsized material..	4	4½	5
Max. size not to exceed 10% of whole..	8	9	10
Capacity—Tons per Hour..	31	47	63
Chain			
Number..	518 F & R	518 F & R	518 F & R
Attachments..	V E-1	V E-1	V E-1
Pitch—Inches..	8	8	8
Working Strength—Lbs..	5225	5225	5225
Buckets—In.			
Length..	20	24	26
Width..	16	18	20
Depth..	8	9	10
Gauge..	10	3/16	3/16
Spacing..	32	32	32
Headshaft—In.			
Diam. of Shaft..	3 7/16	3 15/16	3 15/16
Rev. per minute..	10.7	8.3	8.3
Diam. of Sprocket..	36¼	46¼	46¼
Diam. of Gear..	35.82	48.47	48.47
Pitch of Gear..	1½	1¾	1¾
Face of Gear..	4	5½	5½
1st Countershaft—In.			
Diam. of Pinion..	7.22	7.86	7.86
Face of Pinion..	4½	6	6
Diameter of Shaft..	2 15/16	2 15/16	2 15/16
Rev. per minute..	54	52	52
Diam. of Gear..	25.09	32.00	32.00

No. of Elevator	3260	3261	3262
1st Countershaft—In. Continued			
Pitch of Gear..	1¼	1½	1½
Face of Gear..	3	4	4
2nd Countershaft—Inches			
Diam. of Pinion..	6.01	7.22	7.22
Face of Pinion..	3¼	4½	4½
Diam. of Shaft..	1 15/16	2 7/16	2 7/16
Rev. per minute..	225	230	230
H. P. Required—Max. Ctrs..	8	11	13.5
Corner Shafts—In.			
Diam. Shaft—upper corner..	2 7/16	2 15/16	2 15/16
Diam. Sprocket—upper corner..	26⅜	31⅛	31⅛
Diam. Shaft—lower corner..	1 15/16	2 7/16	2 7/16
Diam. Sprocket—lower corner..	26⅜	26⅜	31⅛
Foot Shaft—In.			
Diam. of Shaft..	1 15/16	2 7/16	2 7/16
Diam. of Sprocket..	26⅜	26⅜	31⅛
Approx. Shipping Weight—Lbs.			
Chain and Buckets per Ft. Ctrs..	58	78	86
Mach. Terminals..	2975	4275	4481
Casing Terminals..	4580	5200	5775
Casing per Foot..	143	151	153
Trough per Foot..	35	39	42

Using No. 518 F. and R. Chain

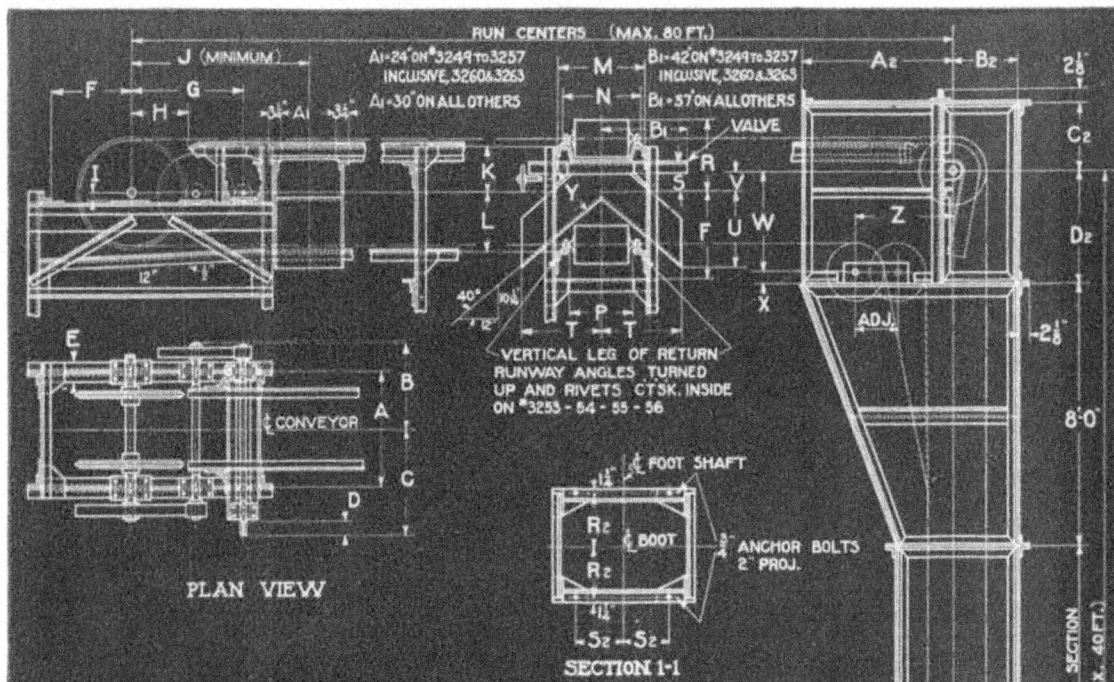

PLAN VIEW

SECTION 1-1

General Dimensions

Dimen- sions	ELEVATOR No.			Dimen- sions	ELEVATOR No.		
	3260	3261	3262		3260	3261	3262
A	40¾	47	49	W	35¼	44⅜	46⅜
B	30⅛	33¼	34¼	X	2¾	3⅛	3⅛
C	39	44	45	Y	12	18	18½
D	6	6	6	Z	37⅜	41⅞	45⅞
E	6¼	8¼	8¼	A2	63⅜	69⅞	73⅞
F	30½	37¼	38¼	B2	26⅝	30⅛	31⅛
G	37⅛	47¾	47¾	C2	24	26½	28½
H	19	22	22	D2	38	47½	49½
I	4	4⅝	4⅝	E2	8'-0"	8'-0"	10'-0"
J	60	75	75	F2	31¾	35¾	37¾
K	16⅛	21¼	21¼	G2	50	52	60
L	20½	26⅜	26⅜	H2	0	2⅜	0
M	30½	34½	36½	J2	7'-8"	7'-8"	9'-4"
N	28½	32½	34½	K2	8½	8½	9½
P	23	27	29	L2	14	16½	18½
R	25⅝	31¾	32¾	M2	24	25½	29½
S	7½	11¾	10¾	N2	23½	24⅜	29
T	25	30	30	P2	26⅝	27¾	31⅛
U	19⅛	25⅛	26⅛	R2	17¾	19¾	20¾
V	5⅛	7⅝	7⅝	S2	19	20	25

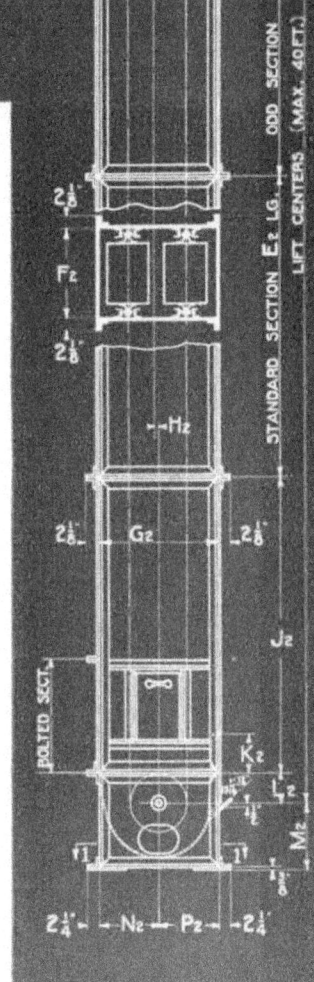

Using No. 276 S. T. R. Chain

Installation View of Jeffrey V-Bucket Conveyor using No. 276 Steel Thimble Roller Chain.

Specifications

MATERIAL: Coal or similar material weighing approximately 50 lbs. per cu. ft.
SPEED: 100 Feet per minute. MAX. CENTERS: 40 ft. vertical, 80 ft. horizontal.

No. of Elevator	3263	3264	3265	No. of Elevator	3263	3264	3265
Size of Material—Inches				**1st Countershaft—**			
Uniform or Average of				**Inches—Continued**			
unsized material	4	4½	5	Rev. per minute	56	52	52
Max. size not to ex-				Diam. of Gear	25.09	32.00	32.00
ceed 10% of whole	8	9	10	Pitch of Gear	1¼	1½	1½
				Face of Gear	3	4	4
Capacity—Tons per							
Hour	28	41	56	**2nd Countershaft—Inches**			
				Diam. of Pinion	6.01	7.22	7.22
Chain				Face of Pinion	3¼	4½	4½
Number	276 STR	276 STR	276 STR	Diam. of Shaft	1⅝	2 7/16	2 7/16
Attachments	Washer	Washer	Washer	Rev. per minute	235	230	230
Pitch—Inches	12	12	12	H. P. Required—			
Working Strength—Lbs.	5200	5200	5200	Max. Ctrs.	6.5	9	11.0
Buckets—In.				**Corner Shaft—In.**			
Length	20	24	26	Diam. Shaft—upper corner	2 7/16	2 15/16	2 15/16
Width	16	18	20	Diam. Sprocket—upper			
Depth	8	9	10	corner	31⅜	31⅜	31⅜
Gauge	10	3/16	3/16	Diam. Shaft—lower corner	1 15/16	2 7/16	2 7/16
Spacing	36	36	36	Diam. Sprocket—lower			
				corner	24	31⅜	31⅜
Headshaft—In.							
Diam. of Shaft	2 15/16	3 15/16	3 15/16	**Foot Shaft—In.**			
Rev. per minute	10.0	8.3	8.3	Diam. of Shaft	1 15/16	2 7/16	2 7/16
Diam. of Sprocket	38⅞	46½	46½	Diam. of Sprocket	24	31⅜	31⅜
Diam. of Gear	40.12	48.47	48.47				
Pitch of Gear	1½	1¾	1¾	**Approx. Shipping**			
Face of Gear	4	5½	5½	**Weight—Lbs.**			
				Chain and Buckets			
1st Countershaft—Inches				per Ft. Ctrs.	74	90	98
Diam. of Pinion	7.22	7.86	7.86	Mach. Terminals	3225	4750	4810
Face of Pinion	4½	6	6	Casing Terminals	4720	5705	5775
Diam. of Shaft	2 7/16	2 15/16	2 15/16	Casing per Foot	135	152	153
				Trough per Foot	42	38	41

Using No. 276 S. T. R. Chain

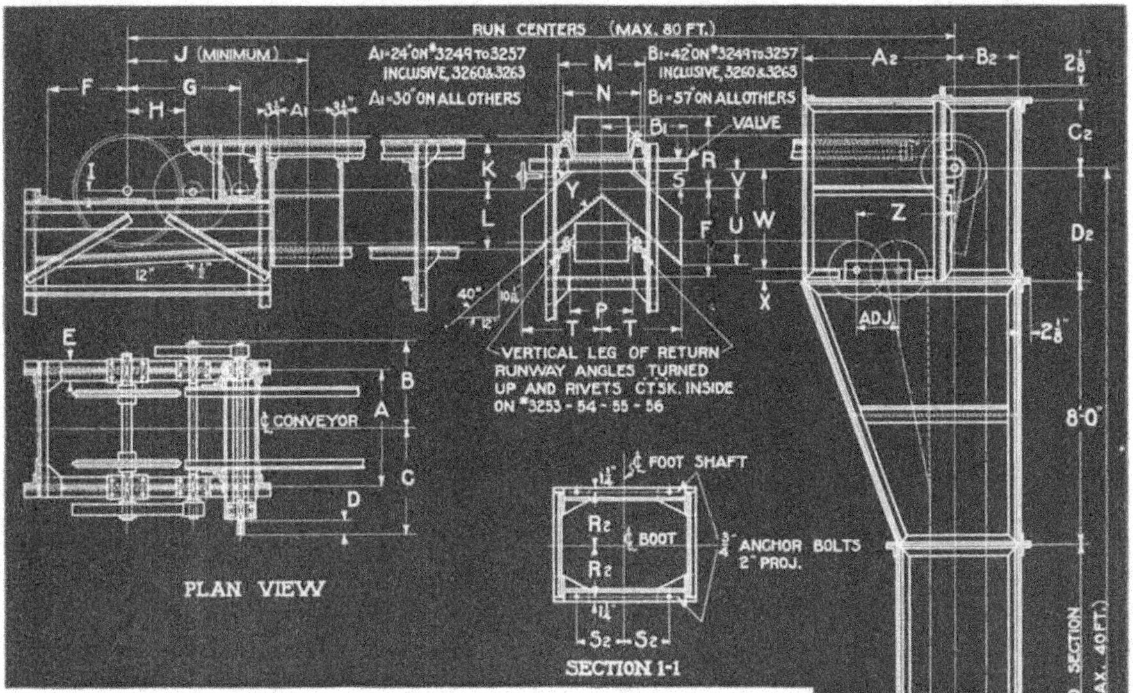

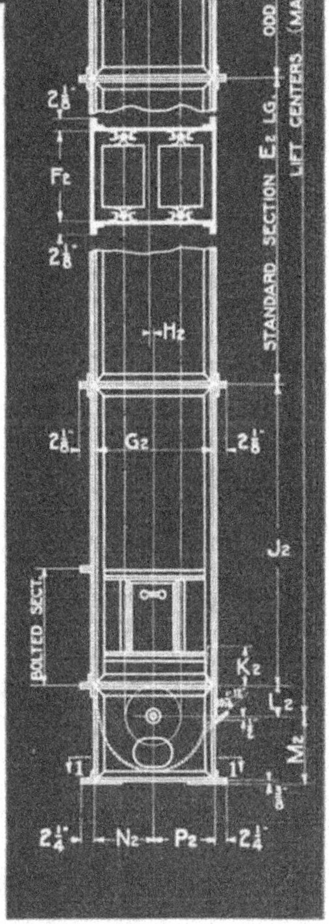

General Dimensions

Dimensions	ELEVATOR No.			Dimensions	ELEVATOR No.		
	3263	3264	3265		3263	3264	3265
A	39	47	49	W	33¼	46⅜	46⅜
B	28¼	33¼	34¼	X	2¾	3⅛	3⅛
C	37	44	45	Y	12	18	18½
D	6	6	6	Z	37⅜	45⅞	45⅞
E	6¼	8¼	8¼	A2	63⅜	73⅞	73⅞
F	31½	37¾	38¾	B2	28⅝	31⅛	31⅛
G	39¼	47¾	47¾	C2	26	28½	28½
H	19	22	22	D2	36	49½	49½
I	3⅝	4⅝	4⅝	E2	10′-0″	10′-0″	10′-0″
J	62	75	75	F2	31¾	35¾	37¾
K	16	19⅝	19⅝	G2	48	60	60
L	23	28	28	H2	3½	0	0
M	30½	34½	36½	J2	10′-0″	9′-4″	9′-4″
N	28½	32½	34½	K2	10½	9½	9½
P	24	28	30	L2	12	18½	18½
R	26⅝	31⅜	32⅜	M2	24	27½	29½
S	8½	11¼	10¼	N2	23	29	29
T	25	30	30	P2	25⅛	31⅛	31⅛
U	18⅛	25⅝	26⅝	R2	17¾	19¾	20¾
V	4	7⅝	7⅝	S2	19	25	25

Using No. 180 S. T. R. Chain

Installation View of Jeffrey V-Bucket Conveyor using No. 180 Steel Thimble Roller Chain.

Specifications

MATERIAL: Coal or similar material weighing approximately 50 lbs. per cu. ft.
SPEED: 100 Feet per minute. MAX. CENTERS: 40 ft. vertical, 80 ft. horizontal.

Number of Elevator	3266	3267	Number of Elevator	3266	3267
Size of Material—Inches			**1st Countershaft—In.–Cont'd**		
Uniform or Average of			Diameter of Shaft	2 13/16	3 14/16
unsized material	5	6	Revolutions per minute	52	46
Maximum size not to exceed			Diameter of Gear	32.00	35.82
10% of whole	10	12	Pitch of Gear	1½	1½
			Face of Gear	4	4
Capacity—Tons per Hour	56	92			
			2nd Countershaft—Inches		
Chain			Diameter of Pinion	7.22	7.22
Number	180 S T R	180 S T R	Face of Pinion	4½	4½
Attachments	Washer	Washer	Diameter of Shaft	2 7/16	2 13/16
Pitch—Inches	12	12	Revolutions per minute	230	230
Working Strength—Lbs	6500	6500	H. P. Required—Max. Ctrs.	11.5	17.5
Buckets—Inches			**Corner Shafts—Inches**		
Length	26	30	Diameter Shaft—upper corner	2 15/16	3 7/16
Width	20	24	Diameter Sprocket—upper		
Depth	10	12	corner	31 3/8	35
Gauge	3/16	¼	Diameter Shaft—lower corner	2 7/16	2 13/16
Spacing	36	36	Diameter Sprocket—lower		
			corner	31 3/8	35
Headshaft—Inches					
Diameter of Shaft	3 15/16	4 15/16	**Foot Shaft—Inches**		
Revolutions per minute	8.3	7.15	Diameter of Shaft	2 7/16	2 13/16
Diameter of Sprocket	46½	54½	Diameter of Sprocket	31 3/8	35
Diameter of Gear	48.47	55.87 C S			
Pitch of Gear	1¾	1½	**Approx. Shipping Wgt.—Lbs.**		
Face of Gear	5½	4½	Chain and Bucket per Ft. Ctrs.	106	140
			Machinery Terminals	4835	6760
1st Countershaft—Inches			Casing Terminals	5775	6460
Diameter of Pinion	7.86	8.92 C S	Casing per Foot	153	180
Face of Pinion	6	4¾	Trough per Foot	41	46

Using No. 180 S. T. R. Chain

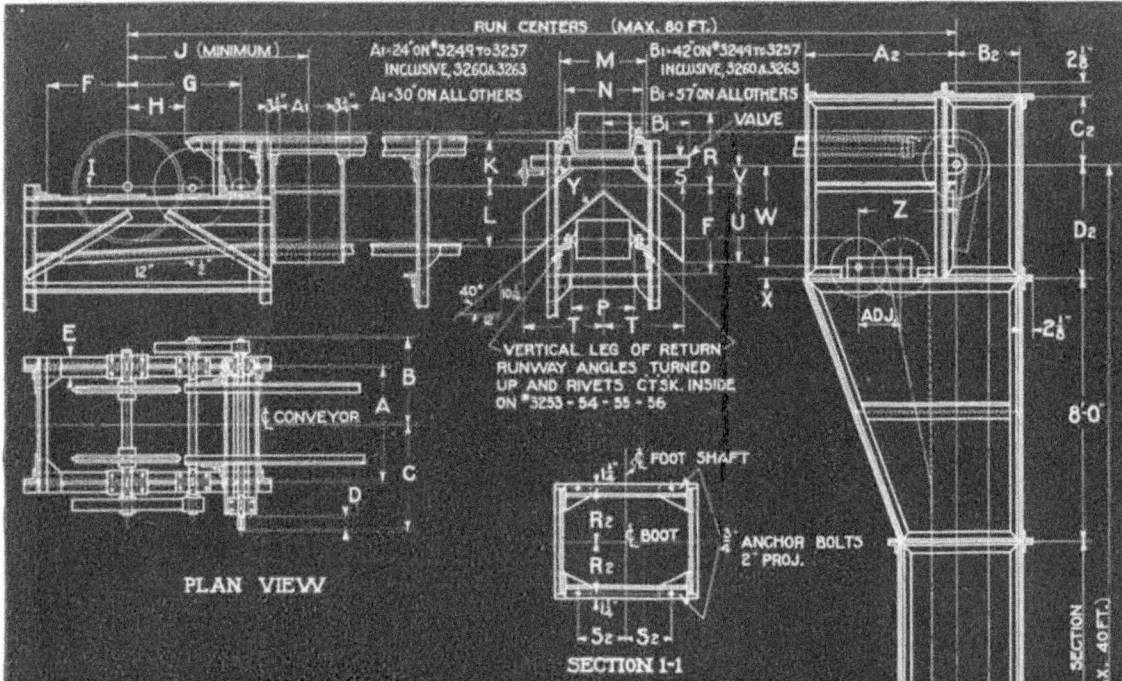

PLAN VIEW

SECTION 1-1

General Dimensions

Dimensions	ELEVATOR No.		Dimensions	ELEVATOR No.	
	3266	3267		3266	3267
A	49¼	57¾	W	46⅜	52¼
B	34⅜	41⅛	X	3⅛	4
C	45	53	Y	18½	20
D	6	8	Z	45⅞	53¼
E	8¼	10¼	A2	73⅞	86¼
F	38¾	43½	B2	31⅛	36¾
G	47¾	54	C2	28½	31¾
H	22	26	D2	49½	56¼
I	4⅝	5⅝	E2	10'-0"	8'-0"
J	75	84	F2	37¾	43¾
K	19⅝	24	G2	60	70
L	28	30¾	H2	0	0
M	37½	41½	J2	9'-4"	7'-8"
N	35½	39½	K2	9½	11½
P	30	34	L2	18½	19¾
R	32⅜	38¾	M2	29½	32¼
S	10¼	12½	N2	29	33⅜
T	30	35	P2	31⅛	36¾
U	26⅝	28½	R2	20¾	23¾
V	7⅝	10	S2	25	30

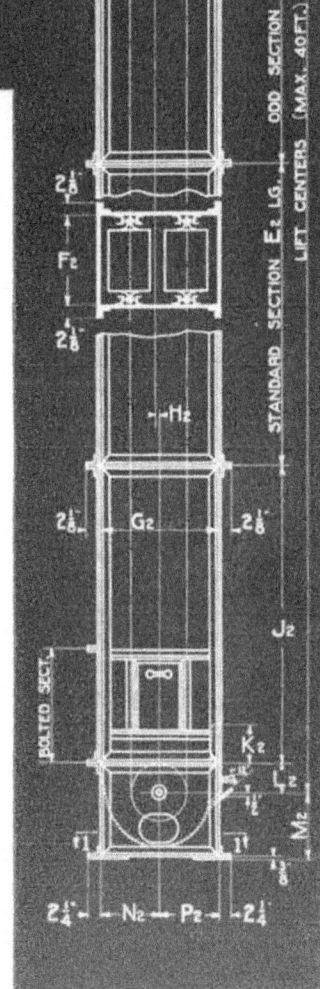

V-Bucket Conveyors
Using No. 182½ S. T. R. Chain

Installation of Jeffrey V-Bucket Conveyor using No. 182½ Steel Thimble Roller Chain.

Specifications

MATERIAL: Coal or similar material weighing approximately 50 lbs. per cu. ft.
SPEED: 100 Feet per minute. MAX. CENTERS: 40 ft. vertical, 80 ft. horizontal.

Number of Elevator	3268	Number of Elevator	3268
Size of Material—Inches		**1st Countershaft—Inches—Continued**	
Uniform or Average of unsized Material	6	Diameter of Shaft	3 13/16
Maximum size not to exceed 10% of whole	12	Revolutions per minute	48
		Diameter of Gear	35.82
Capacity—Tons per Hour	92	Pitch of Gear	1½
		Face of Gear	4
Chain			
Number	182½ S T R	**2nd Countershaft—Inches**	
Attachments	Washer	Diameter of Pinion	7.22
Pitch—Inches	18	Face of Pinion	4½
Working Strength—Lbs.	9700	Diameter of Shaft	2 13/16
		Revolutions per Minute	240
Buckets—Inches		H. P. Required—Max. Ctrs.	18.0
Length	30		
Width	24	**Corner Shafts—Inches**	
Depth	12	Diameter Shaft—upper corner	3 7/16
Gauge	¼	Diameter Sprocket—upper corner	41½
Spacing	36	Diameter Shaft—lower corner	2 13/16
		Diameter Sprocket—lower corner	36
Headshaft—Inches			
Diameter of Shaft	4 13/16	**Foot Shaft—Inches**	
Revolutions per Minute	7.40	Diameter of Shaft	2 13/16
Diameter of Sprocket	52¾	Diameter of Sprocket	36
Diameter of Gear	55.87 C S		
Pitch of Gear	1½	**Approx. Shipping Weight—Lbs.**	
Face of Gear	4½	Chain and Buckets per Ft. Ctrs.	156
		Machinery Terminals	7950
1st Countershaft—Inches		Casing Terminals	6510
Diameter of Pinion	8.92 C S	Casing per Foot	180
Face of Pinion	4¾	Trough per Foot	46

60

Using No. 182½ S. T. R. Chain

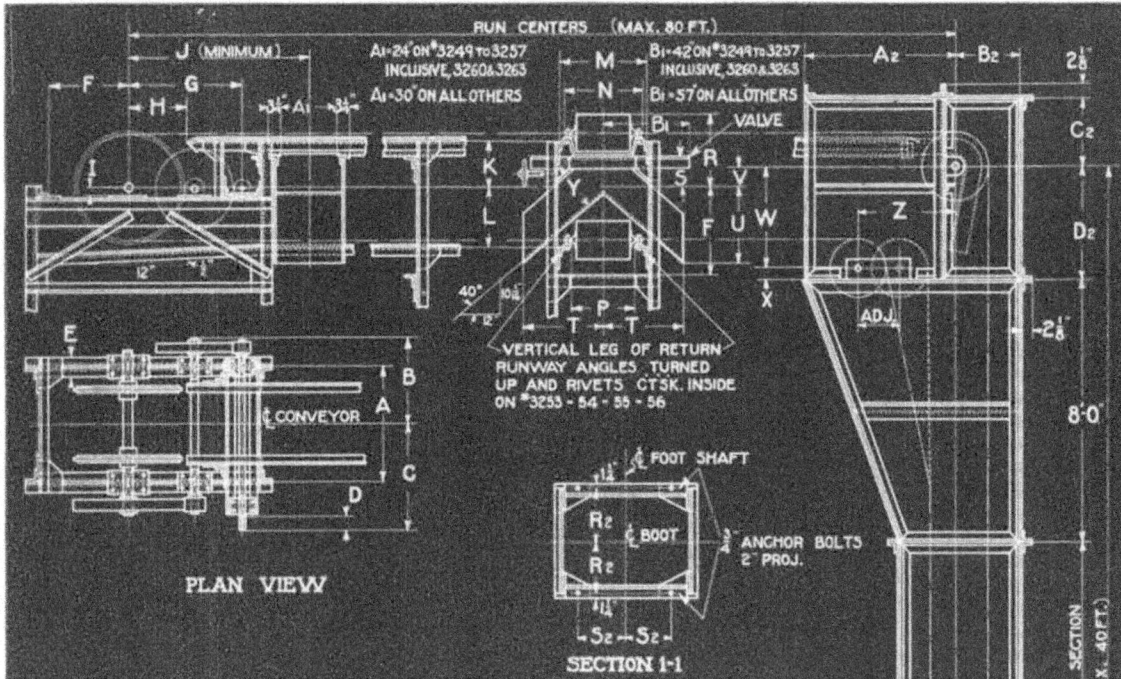

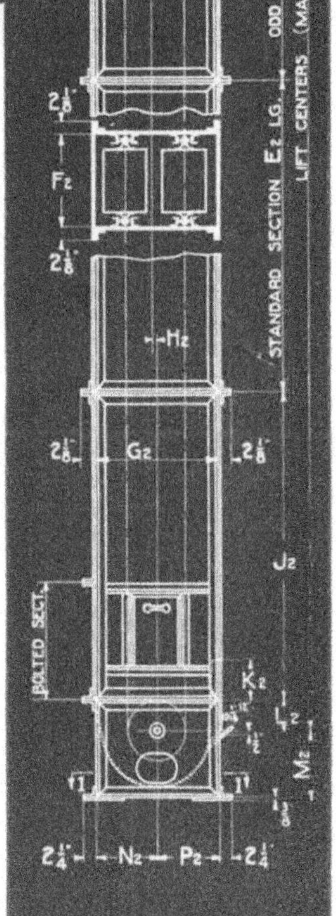

PLAN VIEW

SECTION 1-1

General Dimensions

Dimensions	ELEVATOR No. 3268	Dimensions	ELEVATOR No. 3268
A	58¼	W	48¼
B	41⅜	X	4
C	53	Y	20
D	8	Z	51¼
E	10¼	A2	84¼
F	43¼	B2	39¾
G	54	C2	33¾
H	26	D2	52¼
I	5⅝	E2	8'-0"
J	84	F2	43¾
K	21¾	G2	70
L	31	H2	3
M	42	J2	8'-0"
N	40½	K2	11½
P	34½	L2	19¾
R	37	M2	32¼
S	10¾	N2	33⅜
T	35	P2	36¾
U	30¼	R2	23¾
V	6	S2	30

Bucket
Elevators

Section

3

Seventy-five feet of Jeffrey Standard Elevator handling 30 tons of crushed coal per hour at a large steel plant in the Pittsburgh district. This complete Jeffrey equipment consists of a 12'-0" x 12'-0" steel track hopper, reciprocating plate feeder, single roll coal crusher, elevator and an electrically controlled traveling hopper which distributes the coal to the stokers.

This plant is also equipped with a Jeffrey Standard Ashes Elevator which elevates the ashes from the boiler room floor and discharges them directly into railroad cars.

The Advantages of using Jeffrey Standardized Elevators

JEFFREY Standard Elevators are made vertical or upon an incline and can be furnished with or without steel casings. Their capacities range from 6½ to 36 tons per hour with vertical lifts of from 10 to 75 feet. They consist of endless chains provided with buckets, of steel or malleable iron, spaced at short equal intervals, or close together as indicated on the detailed drawings given throughout this section.

Make quicker delivery possible.

Heretofore, when an elevator was specified, it was necessary to make layouts and complete drawings for that particular elevator; thus entailing considerable expense and much delay. Now all this cost is saved the purchaser by specifying one of the Standard Elevators given in this section. In addition to saving the expense of layouts and drawings, the purchaser is further benefited by a quicker delivery, made possible by the placing of Jeffrey Standard Elevators upon a manufacturing basis. The ease with which a Standard Elevator or other Standard units can be selected to exactly meet your requirements will quickly give this book a place of ready reference in your files.

9951-A

JEFFREY Standard Bucket Elevators have their application to Boiler Houses in the handling of both Coal and Ashes.

Where Coal is to be delivered but a short distance from an elevator to storage space the elevator often can be extended high enough to spout the material to several storage points or boiler hoppers. Thus a Jeffrey Elevator in connection with one or more gravity spouts becomes a maximum of conveying economy.

The drawing at the right shows a typical installation of Jeffrey Standard Bucket Elevator, inclosed in a steel casing. The coal is discharged through track-hopper and fed to Jeffrey Single Roll Crusher by plate feeder, thence to Elevator and Storage Bunkers.

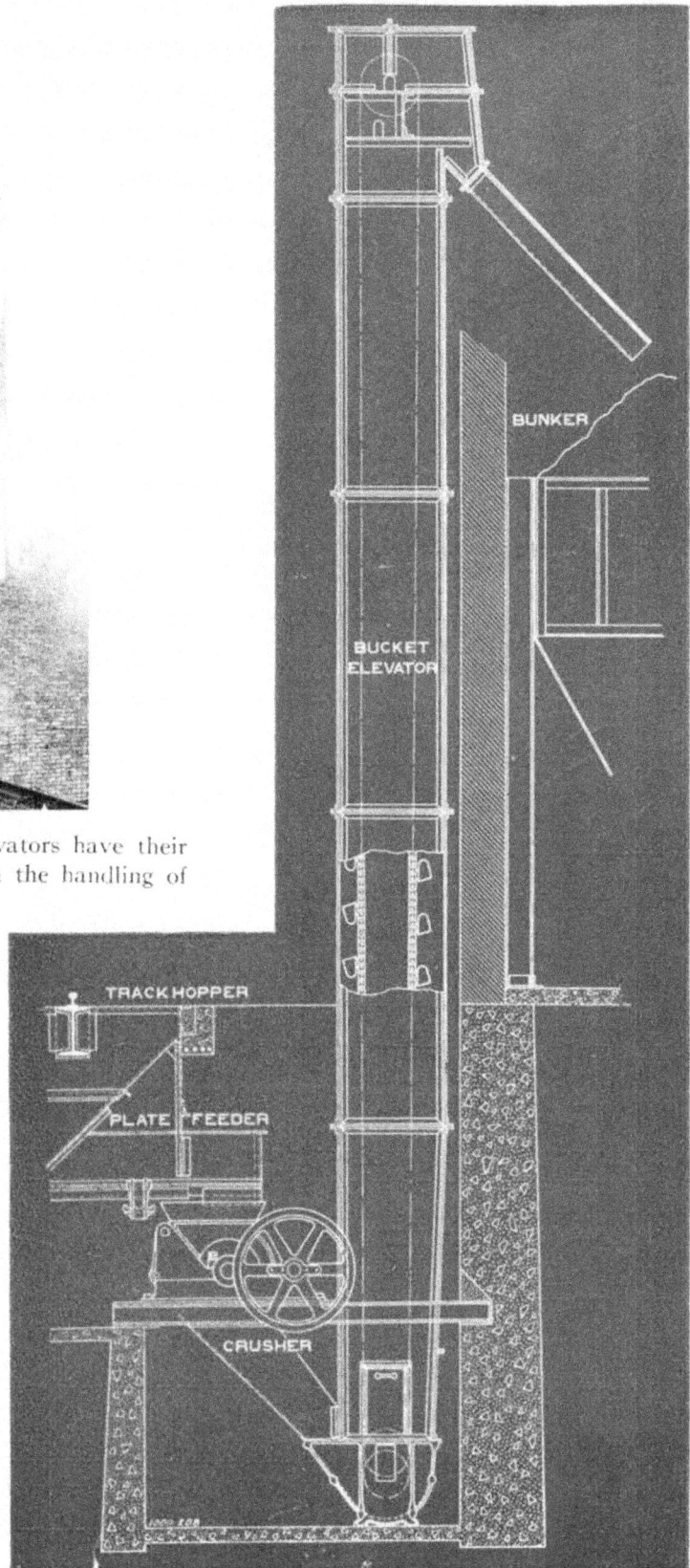

BUNKER

BUCKET ELEVATOR

TRACK HOPPER

PLATE FEEDER

CRUSHER

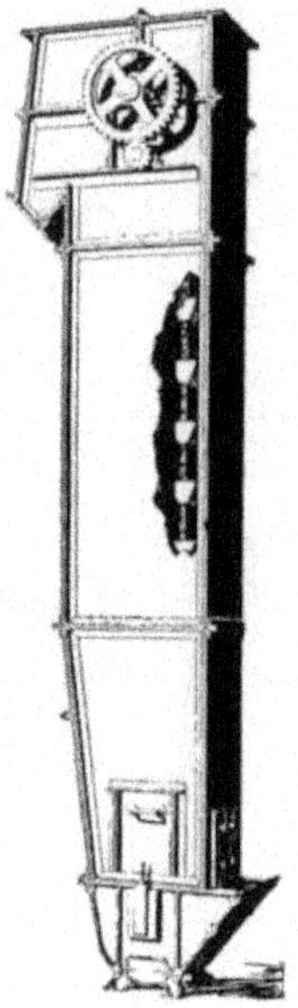

With Casings for Coal and other Similar Materials

Can be furnished without casing if desired.

At the right is shown a typical installation of Jeffrey Bucket Elevator with Steel Casing, handling coal.

Elevator No.	0 to 40 ft. Centers						41 to 80 ft. Centers					
	103	108	111	115	119	122	132	137	140	144	149	152
Maximum Size Piece in Inches												
Not to exceed 10% of whole	2½	3	3½	4	4	4½	2½	3	3½	4	4	4½
Capacity—In Tons per Hour												
Buckets 80% full	6.5	12.2	14	23.2	25	36	6.5	12.2	14	23.2	25	36
Size of Buckets												
Length—Inches	6	8	10	12	14	16	6	8	10	12	14	16
Projection—Inches	4	5	6	7	7	8	4	5	6	7	7	8
Gauge of Steel or	16	14		12	12	10	16	14		12	12	10
Style of Malleable Bucket			A						A			
Spacing—Inches	13	15	24	21	24	24	13	15	24	21	24	24
Chain												
Number and Style	88J	82R	82R	82R	110H	110H	88J	82R	82R	82R	110H	110H
Pitch—Inches	2.61	3.08	3.08	3.08	6	6	2.61	3.08	3.08	3.08	6	6
Attachment	K-1	K-2	K-2	K-2	K-2	K-2	K-1	K-2	K-2	K-2	K-2	K-2
Speed Feet per Minute	192	200	225	200	200	200	192	200	225	200	200	200
Working Strength—Lbs.	815	3000	3000	3000	3900	3900	815	3000	3000	3000	3900	3900
***Horsepower At Countershaft**	.55	1.3	1.3	2.2	2.5	3.3	1.1	2.5	2.6	4.2	4.8	6.6
Head Shaft												
Diameter—Inches	1 11/16	1 11/16	2 7/16	2 7/16	2 15/16	2 15/16	1 11/16	2 7/16	2 15/16	2 15/16	3 7/16	3 7/16
Rev. per Minute	40	40	37.5	33.3	33.3	33.3	40	40	37.5	33.3	33.3	33.3
Diameter Sprocket—Inches	18½	19¾	23½	23½	23½	23½	18½	19¾	23½	23½	23½	23½
Gear Diameter—Inches	23.89	23.89	29.83	29.83	29.83	29.83	23.89	29.83	29.83	29.83	32.00	32.00
Gear Pitch—Inches	1	1	1¼	1¼	1¼	1¼	1	1¼	1¼	1¼	1½	1½
Gear Face—Inches	2½	2½	3	3	3	3	2½	3	3	3	4	4
Countershaft												
Diameter—Inches	1 7/16	1 7/16	1 15/16	1 15/16	2 7/16	2 7/16	1 7/16	1 15/16	2 7/16	2 7/16	2 11/16	2 11/16
Rev. per Minute	188	188	188	167	167	167	188	200	188	167	149	149
Pinion Diameter—Inches	5.12	5.12	6.01	6.01	6.01	6.01	5.12	6.01	6.01	6.01	7.22	7.22
Pinion Face—Inches	2¾	2¾	3¾	3¼	3¼	3¼	2¾	3¾	3¼	3¼	4½	4½
Boot												
Number	111	112	113	113	113	114	111	112	113	113	113	114
Diameter Shaft—Inches	1 3/16	1 7/16	1 15/16	1 15/16	1 15/16	2 7/16	1 7/16	1 7/16	1 15/16	1 15/16	1 15/16	2 7/16
Diameter Sprocket—Inches	11	16¾	17¾	17¾	17¾	21½	11	16¾	17¾	17¾	17¾	21½
Approx. Shipping Weight—Lbs.												
Head and Boot Complete	823	1236	1778	2050	2232	2641	855	1392	1869	2206	2556	2948
Per Foot of intermediate Section Complete with Casing	43	58	70	89	97	106	48	62	76	95	104	112

* Horsepower listed for Maximum Centers.

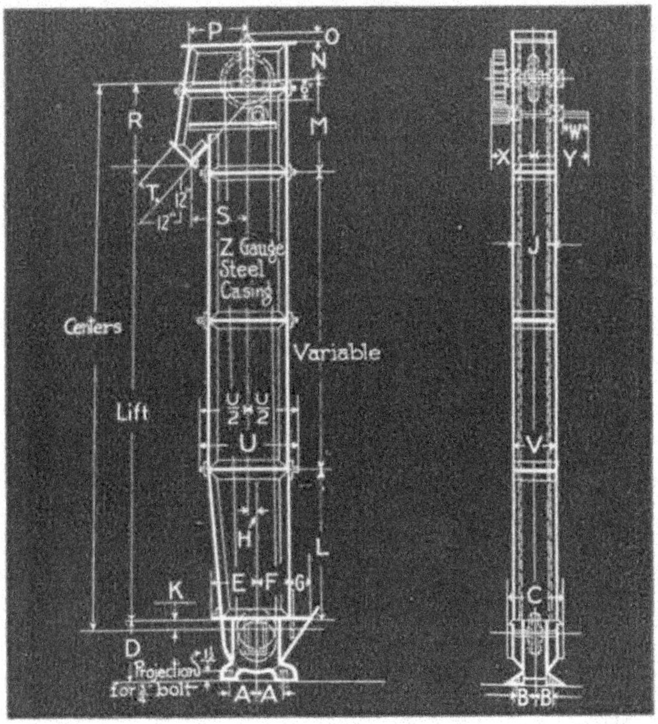

With Casing for Coal and other Similar Materials

Can be furnished without casing if desired.

Always give Elevator Number, Feet Centers, and state whether Gears are to be assembled as shown in drawing or on opposite side of casing.

Elevator No.	0 to 40 ft. Centers						41 to 80 ft. Centers					
	103	108	111	115	119	122	132	137	140	144	149	152
A	6¾	8¼	10	10	10	12	6¾	8¼	10	10	10	12
B	6½	7½	8¾	9¾	10¾	12	7	8	9¼	10¼	11¼	12½
C	18¾	21¾	24¼	26¼	28¼	30¾	19¾	22¾	25¼	27¼	29¼	31¼
D	22¼	26⅞	32 9/16	32 9/16	32 9/16	38¾	22¼	26⅞	32 9/16	32 9/16	32 9/16	38¾
E	15	19	22	22	22	25	15	19	22	22	22	25
F	12	16	19	19	19	22	12	16	19	19	19	22
G	9	10	12	12	12	15	9	10	12	12	12	15
H	4	2	2	3	3	1	5	3	3	5	5	2
J	9	11	13	15	17	19	10	12	14	16	18	20
K	3⅞	3⅝	4 13/16	4 13/16	4 13/16	5¾	3⅞	3⅝	4 13/16	4 13/16	4 13/16	5¾
L	8′-0″	8′-0″	10′-0″	10′-0″	10′-0″	10′-0″	8′-0″	8′-0″	10′-0″	10′-0″	10′-0″	10′-0″
M	32⅛	32⅜	43 3/16	43 3/16	43 3/16	42¼	32⅛	32⅜	43 3/16	43 3/16	43 3/16	42¼
N	16	18	21	22	22	23	17	19	22	24	24	24
O	1⅞	1⅞	1⅞	1⅞	1⅞	1⅞	2⅛	2⅛	2⅛	2⅛	2⅛	2⅛
P	24¾	26	30¾	31¼	31½	31¼	25¾	27¼	32¾	32¾	32¾	32¾
R	27⅝	31	35½	37⅞	37⅞	40 5/16	28⅝	32 1/16	36½	39⅞	39⅞	41 5/16
S	21⅝	25	29½	31⅞	31⅞	34 5/16	22⅝	26 1/16	30½	33⅞	33⅞	35 5/16
T	6	7½	9	10½	10½	12	6	7½	9	10½	10½	12
U	35⅝	39⅝	45¾	47¾	47¾	49¾	38⅛	42⅛	48¼	52¼	52¼	52¼
V	12⅝	14⅝	16¾	18¾	20¾	22¾	14⅛	16⅛	18¼	20¼	22¼	24¼
W	6	6	6	6	6	6	6	6	6	6	7	7
X	13⅝	14⅝	17⅞	18⅞	21⅜	22⅜	14⅛	17⅜	19⅞	20⅞	23⅜	24⅜
Y	14⅝	15⅝	18⅛	19⅛	21⅜	22⅜	15⅛	17⅝	19⅞	20⅞	23¾	24¾
Z	16	16	14	12	12	12	16	16	14	12	12	12

Dimensions given above are in inches except as otherwise noted.

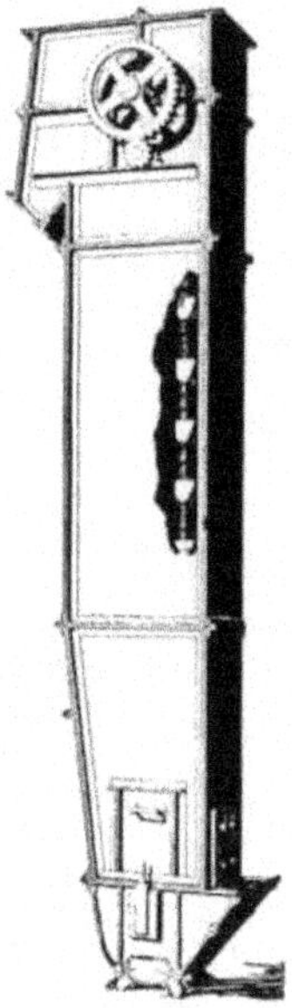

With Casing for Ashes, Coke and other Similar Materials

At the right is shown a typical installation of Jeffrey Bucket Elevator with Steel Casing, handling Ashes.

Elevator No.	0 to 40 ft. Centers				41 to 80 ft. Centers			
	166	167	169	170	179	180	182	183
Maximum Size Piece in Inches								
Not to exceed 10% of whole	4	4	4	4	4	4	4	4
Capacity—In Tons per Hour Ashes	17	17	21	21	17	17	21	21
Buckets 80% full Coke	13	13	16	16	13	13	16	16
Size of Bucket								
Length—Inches	12	12	14	14	12	12	14	14
Projection—Inches	7	7	7	7	7	7	7	7
Style of Malleable Bucket	A	AA	A	AA	A	AA	A	AA
Spacing—Inches	24	24	24	24	24	24	24	24
Chain								
Number and Style	110H	825P	110H	825P	111H	844P	111H	844P
Pitch—Inches	6	4	6	4	4.78	6	4.78	6
Attachment	K-2	K-2	K-2	K-2	K-2	K-2	K-2	K-2
Speed Feet per Minute	200	200	200	200	200	200	200	200
Working Strength—Lbs.	3900	5075	3900	5075	5600	7750	5600	7750
***Horsepower At Countershaft**	2	2	2.5	2.2	3.6	3.6	4.8	4.2
Head Shaft								
Diameter—Inches	2 7/16	2 7/16	2 15/16	2 15/16	2 15/16	2 15/16	3 7/16	3 7/16
Rev. per Minute	33.3	33.3	33.3	33.3	33.3	33.3	33.3	33.3
Diameter Sprocket—Inches	23½	23¼	23½	23½	23	23½	23	23½
Gear Diameter—Inches	29.83	29.83	29.83	29.83	32.00	32.00	32.00	32.00
Gear Pitch—Inches	1¾	1¾	1¾	1¾	1½	1½	1½	1½
Gear Face—Inches	3	3	3	3	4	4	4	4
Countershaft								
Diameter—Inches	1 15/16	1 15/16	2 7/16	2 7/16	2 7/16	2 7/16	2 11/16	2 11/16
Rev. per Minute	168	168	168	168	149	149	149	149
Pinion Diameter—Inches	6.01	6.01	6.01	6.01	7.22	7.22	7.22	7.22
Pinion Face—Inches	3¾	3¾	3¾	3¾	4½	4½	4½	4½
Boot								
Number	113	113	113	113	113	113	113	113
Diameter Shaft—Inches	1 15/16	1 15/16	1 15/16	1 15/16	1 15/16	1 15/16	1 15/16	1 15/16
Diameter Sprocket—Inches	18	17¼	17¾	18	17¾	17¾	17⅝	17¾
Approx. Shipping Weight—Lbs.								
Head and Boot Complete	2207	2253	2339	2427	2451	2478	2658	2672
Per Foot of intermediate Section Complete with Casing	97	101	101	103	104	107	109	110

* Horsepower listed for Maximum Centers.

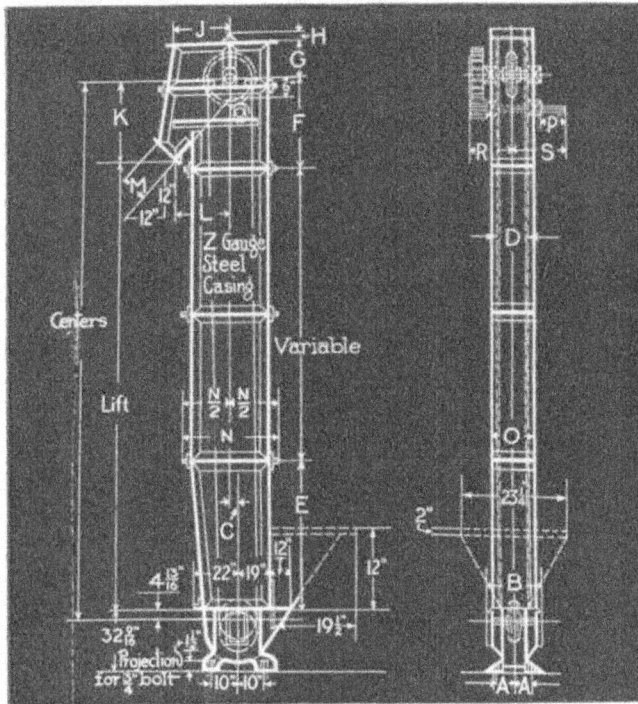

With Casing for Ashes, Coke and other Similar Materials

Always give Elevator Number, Feet Centers, and state whether Gears are to be assembled as shown in drawing or on opposite side of casing.

Elevator No.	0 to 40 ft. Centers				41 to 80 ft. Centers			
	166	167	169	170	179	180	182	183
A	9¾	9¾	10¾	10¾	10¼	10¼	11¼	11¼
B	26¼	26¼	28¼	28¼	27¼	27¼	29¼	29¼
C	3½	3½	3½	3½	4½	4½	4½	4½
D	15	15	17	17	16	16	18	18
E	10'-0"	10'-0"	10'-0"	10'-0"	10'-0"	10'-0"	10'-0"	10'-0"
F	43³⁄₁₆	43³⁄₁₆	43³⁄₁₆	43³⁄₁₆	43³⁄₁₆	43³⁄₁₆	43³⁄₁₆	43³⁄₁₆
G	22½	22½	22½	22½	23½	23½	23½	23½
H	1⅞	1⅞	1⅞	1⅞	2⅛	2⅛	2⅛	2⅛
J	32½	32½	32½	32½	32¾	32¾	32¾	32¾
K	38⅜	38⅜	38⅜	38⅜	39⅜	39⅜	39⅜	39⅜
L	32⅜	32⅜	32⅜	32⅜	33⅜	33⅜	33⅜	33⅜
M	10½	10½	10½	10½	10½	10½	10½	10½
N	48¾	48¾	48¾	48¾	51¼	51¼	51¼	51¼
O	18¾	18¾	20¾	20¾	20¼	20¼	22¼	22¼
P	6	6	6	6	6	6	7	7
R	18⅞	18⅞	21⅜	21⅜	20⅞	20⅞	23⅜	23⅜
S	19⅛	19⅛	21⅜	21⅜	20⅞	20⅞	23¾	23¾
Z	12	12	12	12	12	12	12	12

Dimensions given above are in inches except as otherwise noted.

Power House
Weigh Larries

Section

4

Jeffrey Hand Propelled Weigh Larry

THE Traveling Weigh Larry affords an economical and convenient method of distributing coal from bunkers to stoker magazines. With this outfit the boiler house attendant may accurately weigh and keep a record of all coal delivered to each one of the boilers. The type shown above is operated from floor by chain. The valve is also controlled in the same manner. See opposite page for general dimensions.

General Dimensions of Hand Propelled Weigh Larries

1000 LBS. CAPACITY										2000 LBS. CAPACITY									
Angle of Chute A	B	C	D	F	G	H	J	K	L	Angle of Chute A	B	C	D	F	G	H	J	K	L
60°	23 1/16"	7'- 0 3/4"	8'-0"	3'-6"	4'-8 1/2"					60°	2'-10 1/4"	7'-11"	9'-0"	4'-0"	5'-2 1/2"				
57°-30'	22 3/4"	6'- 5 3/4"	7'-6"	3'-6 3/8"	4'-8 1/2"					57°-30'	2'- 9 13/16"	7'- 4"	8'-6"	4'-0 3/4"	5'-3"				
55°	22 1/8"	5'-10 3/4"	7'-0"	3'-6 1/4"	4'-7 1/2"					55°	2'- 9 5/8"	6'- 8 5/8"	8'-0"	4'-1 3/4"	5'-2 1/2"				
52°-30'	22"	5'- 6 3/4"	6'-9"	3'-7 1/2"	4'-8"					52°-30'	2'- 9 3/16"	6'- 4 1/4"	7'-9"	4'-2 7/8"	5'-3 1/2"				
50°	21 9/16"	5'- 2 1/2"	6'-6"	3'-8 1/2"	4'-8 1/4"	2'-6"	6"	2'-4 1/2"	2'-1"	50°	2'- 8 3/4"	5'-11 5/8"	7'-6"	4'-4 1/4"	5'-4 1/2"	3'-0"	3"	2'-1 1/2"	1'-10"
47°-30'	21"	4'-10 3/4"	6'-3"	3'-9 1/4"	4'-9"					47°-30'	2'- 8 5/16"	5'- 7"	7'-3"	4'-5 1/4"	5'-5"				
45°	20 3/8"	4'- 6"	6'-0"	3'-9 1/2"	4'-9"					45°	2'- 7 7/8"	5'- 2 1/2"	7'-0"	4'-6"	5'-5"				
42°-30'	19 3/4"	4'- 2"	5'-9"	3'-9 5/8"	4' 8 1/2"					42°-30'	2'- 6 13/16"	4'-10 3/8"	6'-9"	4'-6 1/2"	5'-5 1/2"				
40°	18 13/16"	3'-10"	5'-6"	3'-9 1/2"	4'-8"					40°	2'- 6 5/8"	4'- 5 3/4"	6'-6"	4'-6 5/8"	5'-5"				

Jeffrey Motor Propelled Weigh Larry

THE Motor Propelled Weigh Larry is recommended for Boiler Rooms of moderate capacity, where the run is quite long and the number of boilers to be served greater than efficiency and economy would dictate for the hand propelled type. In both the hand propelled and motor propelled types, the scales can be read from the floor. For general dimensions see table on opposite page.

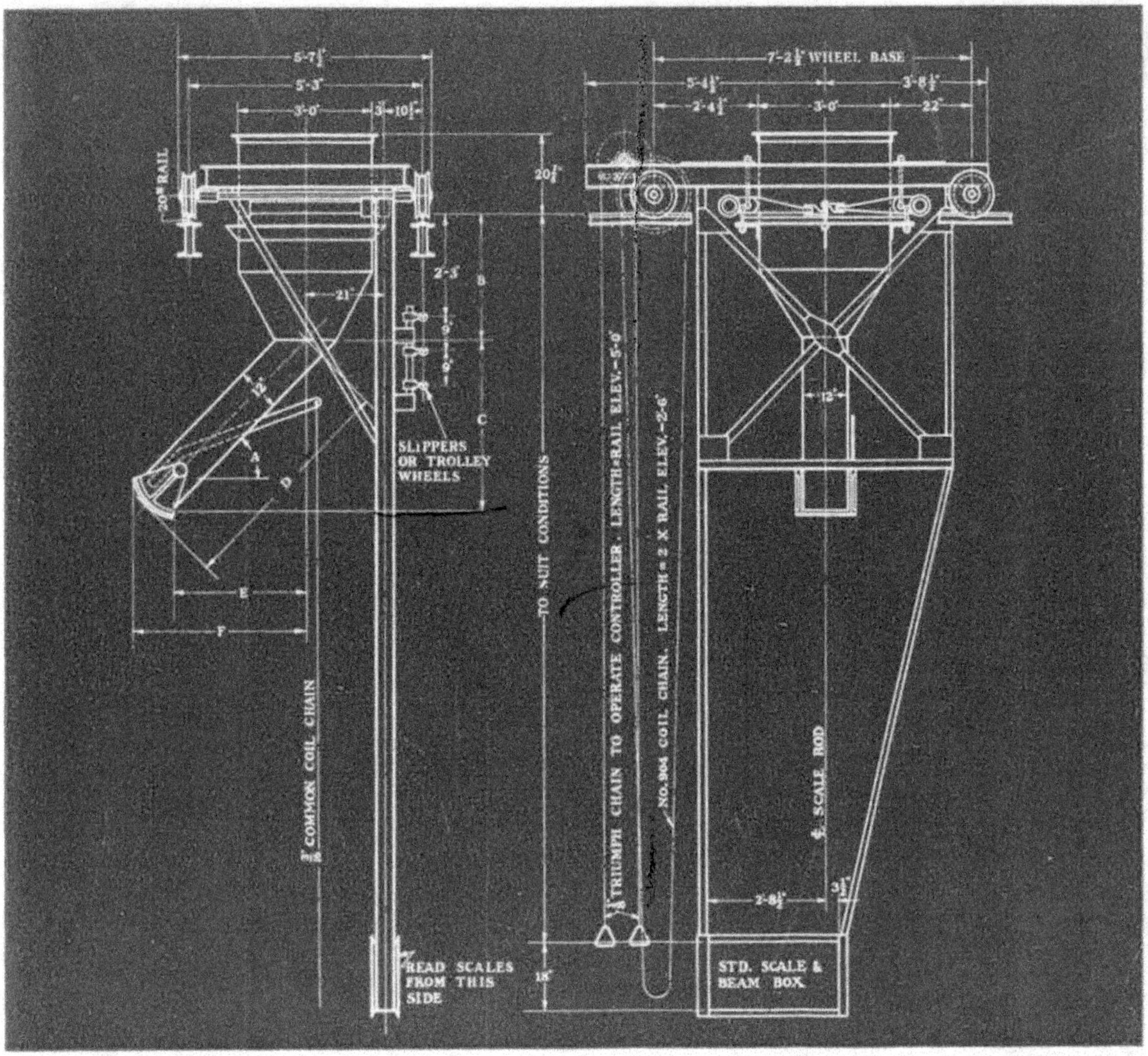

General Dimensions of Motor Propelled Weigh Larry

Capacity 2000 Pounds

Angle of Chute A	B	C	D	E	F	Angle of Chute A	B	C	D	E	F
60°	2'-10¼"	7'-11"	9'-0"	4'-0"	5'-2½"	47°-30'	2'- 8 3/16"	5'- 7"	7'-3"	4'-5¼"	5'-5"
57°-30'	2'- 9 15/16"	7'- 4"	8'-6"	4'-0¾"	5'-3"	45°	2'- 7 9/16"	5'- 2½"	7'-0"	4'-6"	5'-5"
55°	2'- 9⅝"	6'- 8⅝"	8'-0"	4'-1¼"	5'-2½"	42°-30'	2'- 6 15/16"	4'-10⅛"	6'-9"	4'-6½"	5'-5½"
52°-30'	2'- 9 3/16"	6'- 4¼"	7'-9"	4'-2⅞"	5'-3½"	40°	2'- 6⅛"	4'- 5¾"	6'-6"	4'-6⅝"	5'-5"
50°	2'- 8¾"	5'-11⅝"	7'-6"	4'-4¼"	5'-4½"						

Jeffrey Motor Propelled Weigh Larry with Cab

THIS type of Weigh Larry is designed to meet the requirements of large modern Power Houses where many boilers are to be served. The carriage valves and weighing mechanism are all controlled from the operator's cab. The Swivel Spout enables it to serve boilers located on both sides of runway. See opposite page for general dimensions.

General Dimensions of Motor Propelled Weigh Larry with Cab
Capacity 2000 Pounds.

Track Hoppers
Plate Feeders
and Bin Valves

Section

5

Jeffrey Standard Track Hopper with concrete slopes which permit of a quick clean-up.
For general dimensions of standard Track Hoppers see following pages.

A Jeffrey Double Track Hopper serving a railroad Coaling Station. Coal can be easily and quickly handled
from hopper bottom cars by dumping into Jeffrey Track Hopper from which it is delivered to elevating or
conveying equipment by either a plate or apron feeder.

Jeffrey Reciprocating Plate Feeder for regulating the flow of material from Track Hopper to Crusher. This type of feeder is recommended where local conditions will permit of placing the elevator close to railroad track, or a conveyor passing under the track. See following pages for arrangement and dimensions.

A typical installation showing Jeffrey Apron Feeder operating from under the Track Hopper to Crusher. The Apron Feeder works out to advantage where the elevator is somewhat removed from Track Hopper or where it is desired to save in depth of the elevator pit. See following pages for arrangement and dimensions.

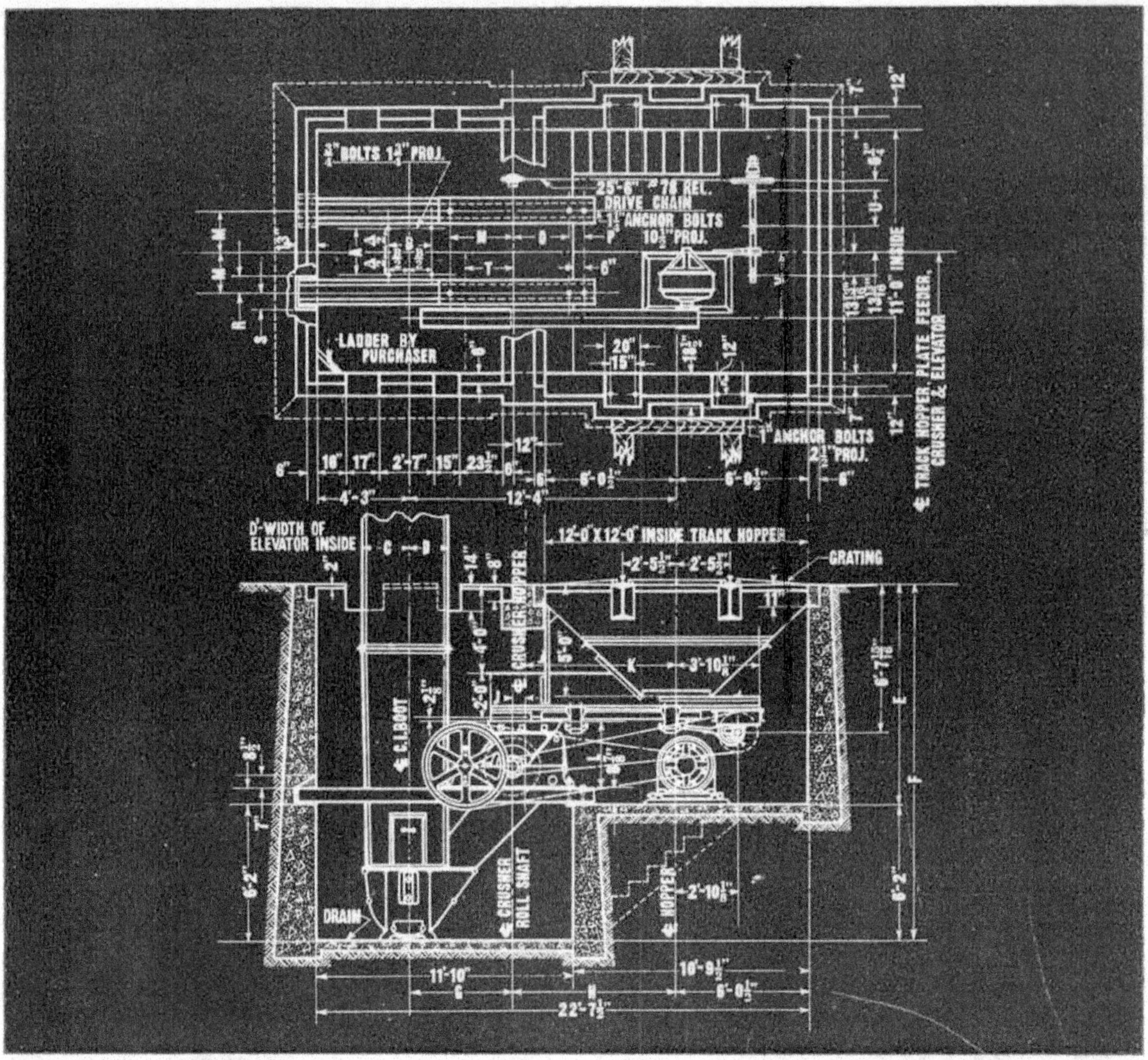

General arrangement of a Jeffrey Standard 12 x 12 Steel Track Hopper with Reciprocating Plate Feeder to handle Coal through Single Roll Crusher to Bucket Elevator.

Capacity Tons per Hour	ELEVATORS						
	No.	Centers	A	B	C	D	D1
23.2	115 144	0'-40' 41'-80'	19¼" 20¼"	20" 20"	2'-1" 2'-5"	19" 19"	15" 16"
25	119 149	0'-40' 41'-80'	21½" 22½"	20" 20"	2'-1" 2'-5"	19" 19"	17" 18"
36	122 152	0'-40' 41'-80'	2'-0" 2'-1"	2'-0" 2'-0"	2'-0" 2'-2"	22" 22"	19" 20"
50	Cont. Bucket	0'-40' 41'-80'	2'-0" 2'-1"	2'-0" 2'-0"	2'-0" 2'-2"	22" 22"	19" 20"

Where local conditions will not permit of installing a standard 12 x 12 Hopper, an 8 x 8 or 10 x 10 Hopper can be furnished.

Full information on request.

‡Grating omitted when maximum size of piece does not exceed tabulated dimension.

Purchaser to determine thickness of pit walls to suit local soil conditions.

For detailed information on Bucket Elevators, see pages 64 to 69.

For detailed information on Single Roll Crushers, see pages 185 to 194.

Size Crusher	Motor Req'd		E	F	G	H	J	K	L	M	N	O	P	R	S	T	U	V	Mesh of Grating ‡
	HP	Speed																	
24x24	25	860	9'-7½"	15'-9½"	4'-8⅜"	7'-7⅞"	5⅜"	7'-2½"	2'-8½"	17⅞"	2'-5½"	2'-2"	14⅞"	6"	8"	21"	12 11/16"	2'-4¾"	14"
30x30	35	860	10'-0"	16'-2"	4'-9½"	7'-6½"	6½"	7'-0"	3'-1"	22⅛"	2'-10¼"	2'-6¾"	8¾"	9"	9"	2'-1½"	18 11/16"	2'-11⅜"	20"

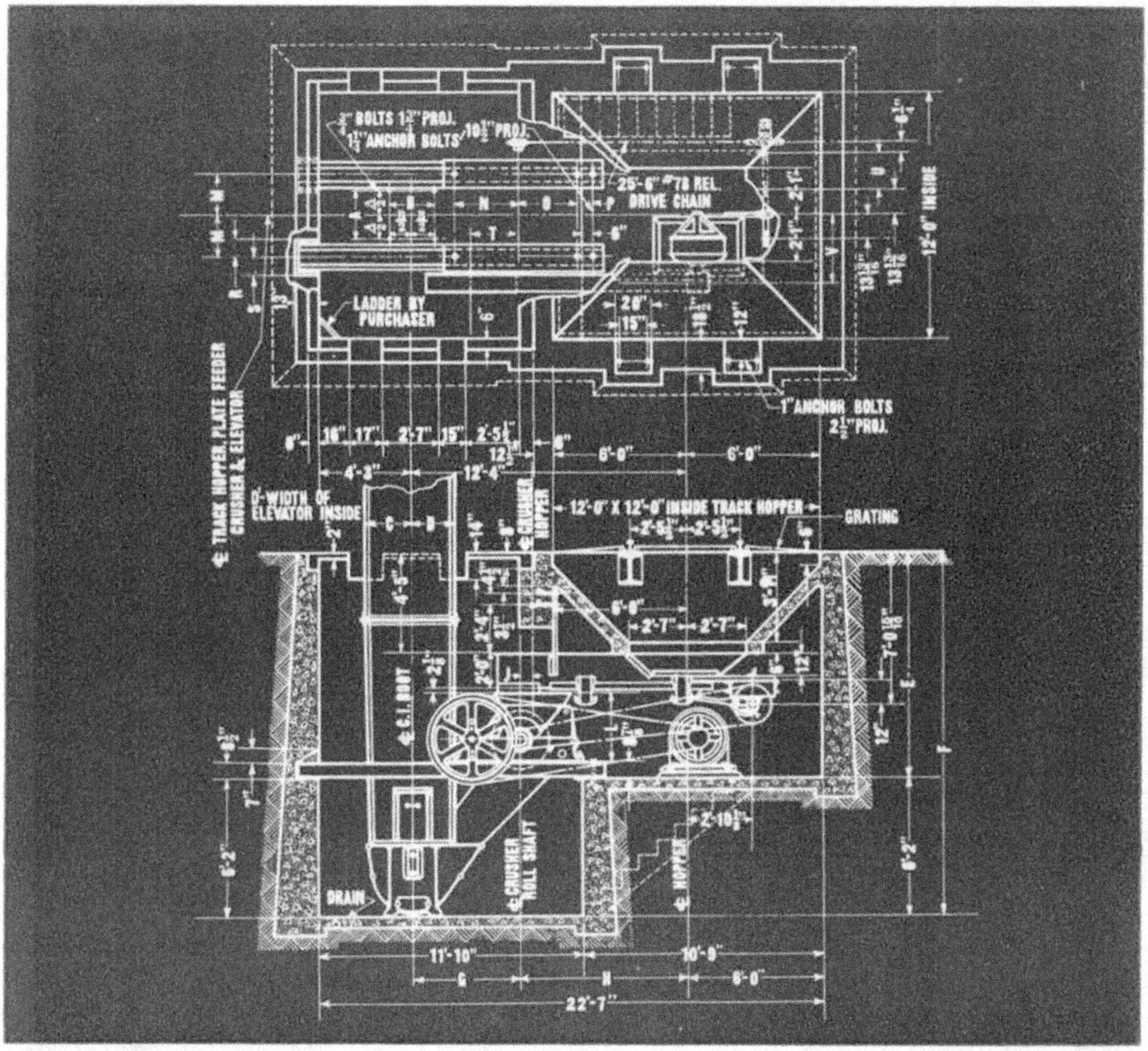

General arrangement of a Jeffrey Standard 12 x 12 Concrete Track Hopper with Reciprocating Plate Feeder to handle Coal through Single Roll Crusher to Bucket Elevator.

ELEVATORS							
Capacity Tons per Hour	No.	Centers	A	B	C	D	D1
23.2	115	0'-40'	19½"	20"	2'-1"	19"	15"
	144	41'-80'	20½"	20"	2'-5"	19"	16"
25	119	0'-40'	21¼"	20"	2'-1"	19"	17"
	149	41'-80'	22½"	20"	2'-5"	19"	18"
36	122	0'-40'	2'-0"	2'-0"	2'-0"	22"	19"
	152	41'-80'	2'-1"	2'-0"	2'-2"	22"	20"
50	Cont. Bucket	0'-40'	2'-0"	2'-0"	2'-0"	22"	19"
		41'-80'	2'-1"	2'-0"	2'-2"	22"	20"

Where local conditions will not permit of installing a standard 12 x 12 Hopper, an 8 x 8 or 10 x 10 Hopper can be furnished.

Full information on request.

‡Grating omitted when maximum size of piece does not exceed tabulated dimensions.

Purchaser to determine thickness of pit walls to suit local soil conditions.

For detailed information on Bucket Elevators, see pages 64 to 69.

For detailed information on Single Roll Crusher, see pages 185 to 194.

Size Crusher	Motor Req'd		E	F	G	H	J	L	M	N	O	P	R	S	T	U	V	‡Mesh of Grating
	HP	Speed																
24x24	25	860	10'-0½"	16'-2½"	4'-8⅛"	7'-7⅝"	5⅜"	2'-8½"	17⅞"	2'-5½"	2'-2"	14⅞"	6"	8"	21"	12¹³⁄₁₆"	2'-4¾"	14"
30x30	35	860	10'-5"	16'-7"	4'-9½"	7'-6½"	6½"	3'-1"	22⅛"	2'-10¾"	2'-6¾"	8¾"	9"	9"	2'-1½"	18¹³⁄₁₆"	2'-11⅜"	20"

83

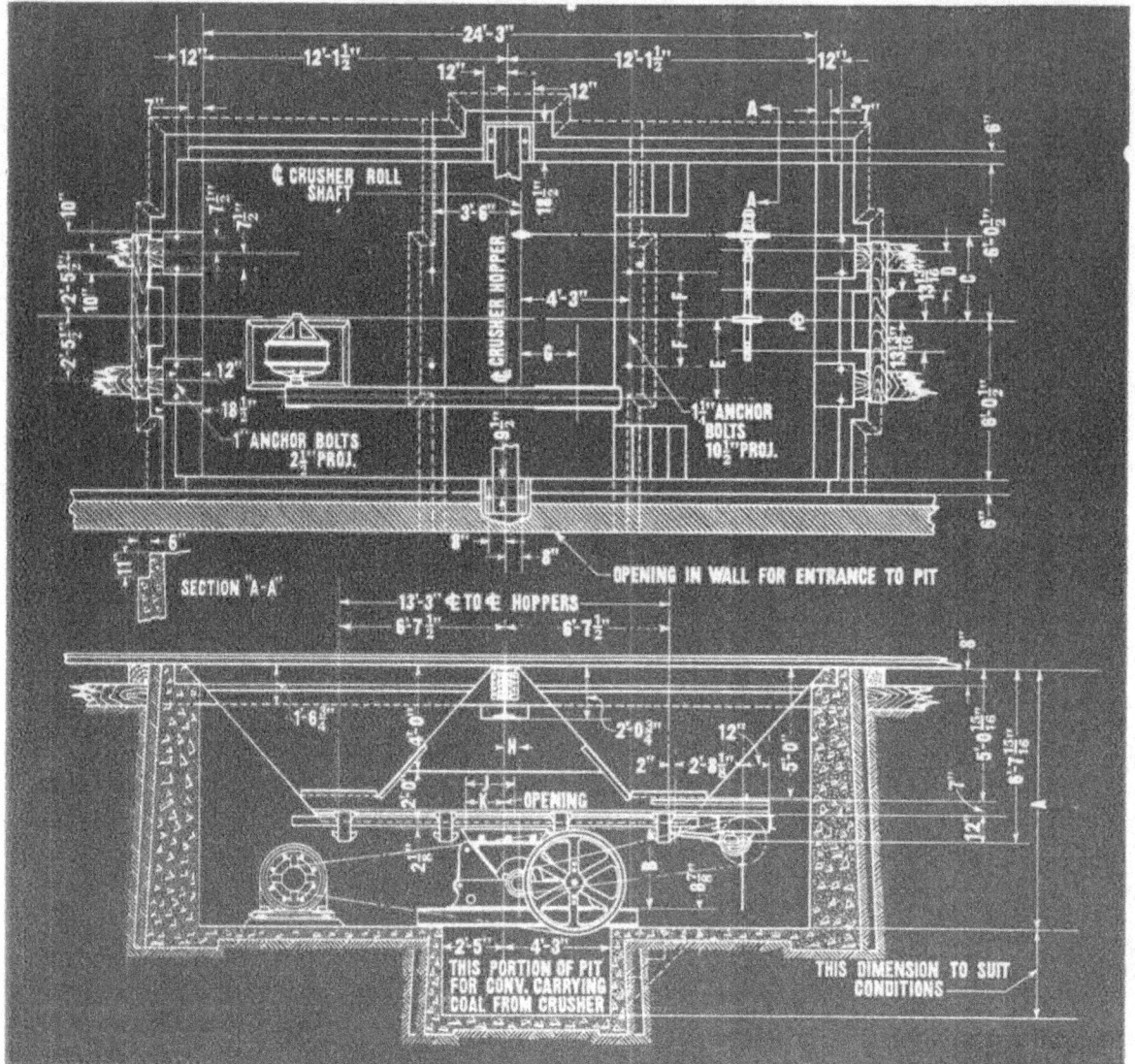

General arrangement of a Jeffrey Double Steel Track Hopper
with double Reciprocating Plate Feeder to handle Coal
through Single Roll Crusher to Main Conveyor.

Size Crusher	A	B	C	D	E	F	G	H	J	K	Mesh of Grating ‡	Motor Req'd	
												H. P.	Speed
24 x 24	9'-7½"	2'-8½"	2'-9"	12¹⁵⁄₁₆"	2'- 4¾"	19¼"	21"	5⅜"	2'-0"	15"	14"	25	860
30 x 30	10'-0"	3'-1"	3'-3"	18¹⁵⁄₁₆"	2'-11⅜"	22⅛"	2'-1½"	6½"	2'-6"	18"	20"	35	860

‡Grating omitted when maximum size of piece does not exceed tabulated dimensions.
Purchaser to determine thickness of pit walls to suit local soil conditions.
For detailed information on Single Roll Crusher, see pages 185 to 194.

Where local conditions will not permit of installing a Standard 12x12 Hopper, an 8x8 or 10x10 Hopper can be furnished. Full information upon request.

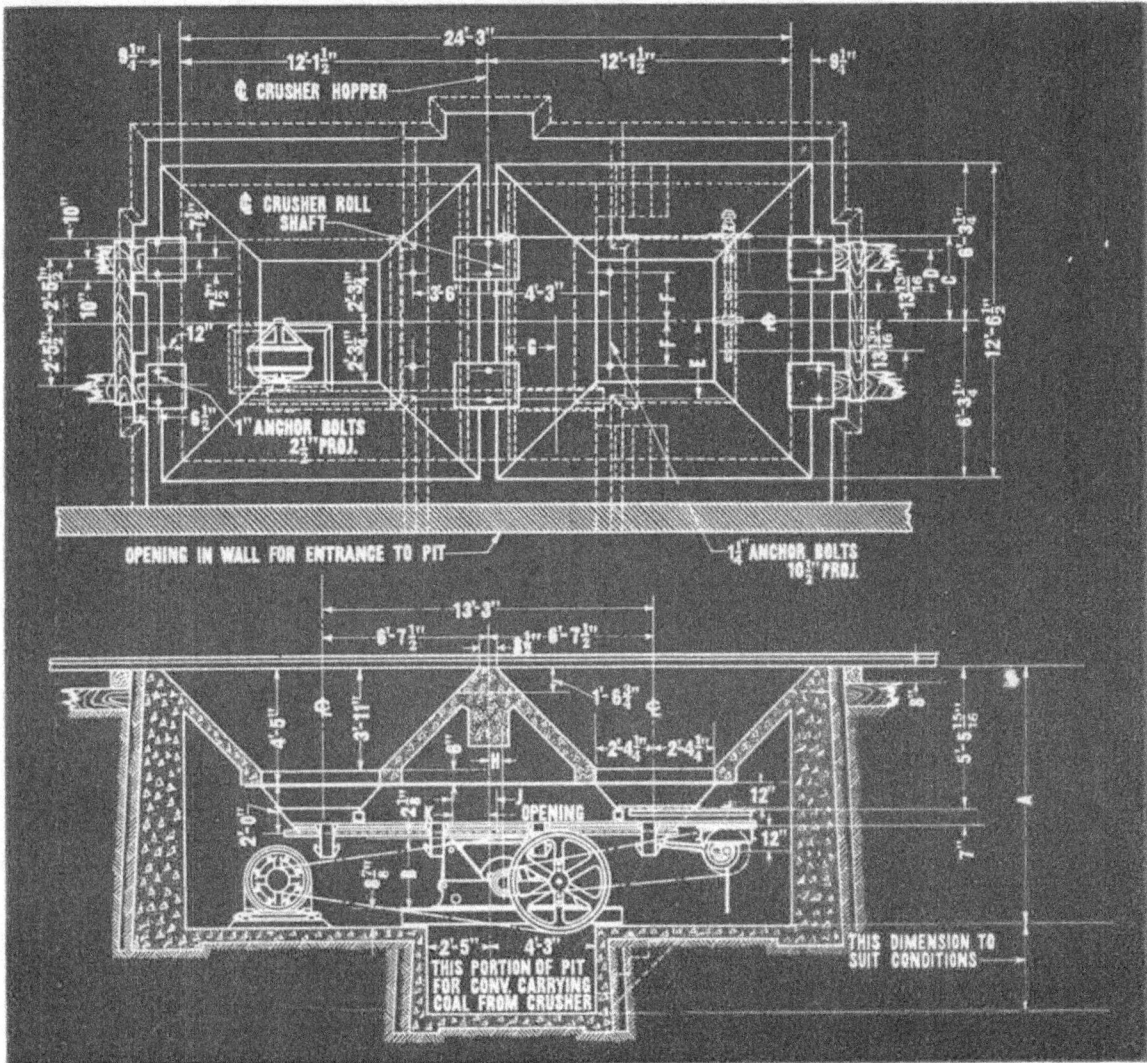

General arrangement of a Jeffrey Double Concrete Track Hopper
with double Reciprocating Plate Feeder to handle Coal
through Single Roll Crusher to Main Conveyor.

Size Crusher	A	B	C	D	E	F	G	H	J	K	Mesh of Grating ‡	Motor Req'd	
												H. P.	Speed
24 x 24	10'-0½"	2'-8½"	2'-9"	12¹⁵⁄₁₆"	2'- 4¾"	19¼"	21"	5⅜"	2'-0"	15"	14"	25	860
30 x 30	10'-5"	3'-1"	3'-3"	18¹⁵⁄₁₆"	2'-11⅜"	22⅜"	2'-1½"	6½"	2'-6"	18"	20"	35	860

‡Grating omitted when maximum size of piece does not exceed tabulated dimensions.
Purchaser to determine thickness of pit walls to suit local soil conditions.
For detailed information on Single Roll Crusher, see pages 185 to 194

Where local conditions will not permit of installing a Standard 12x12 Hopper, an 8x8 or 10x10 Hopper can be furnished. Full information upon request.

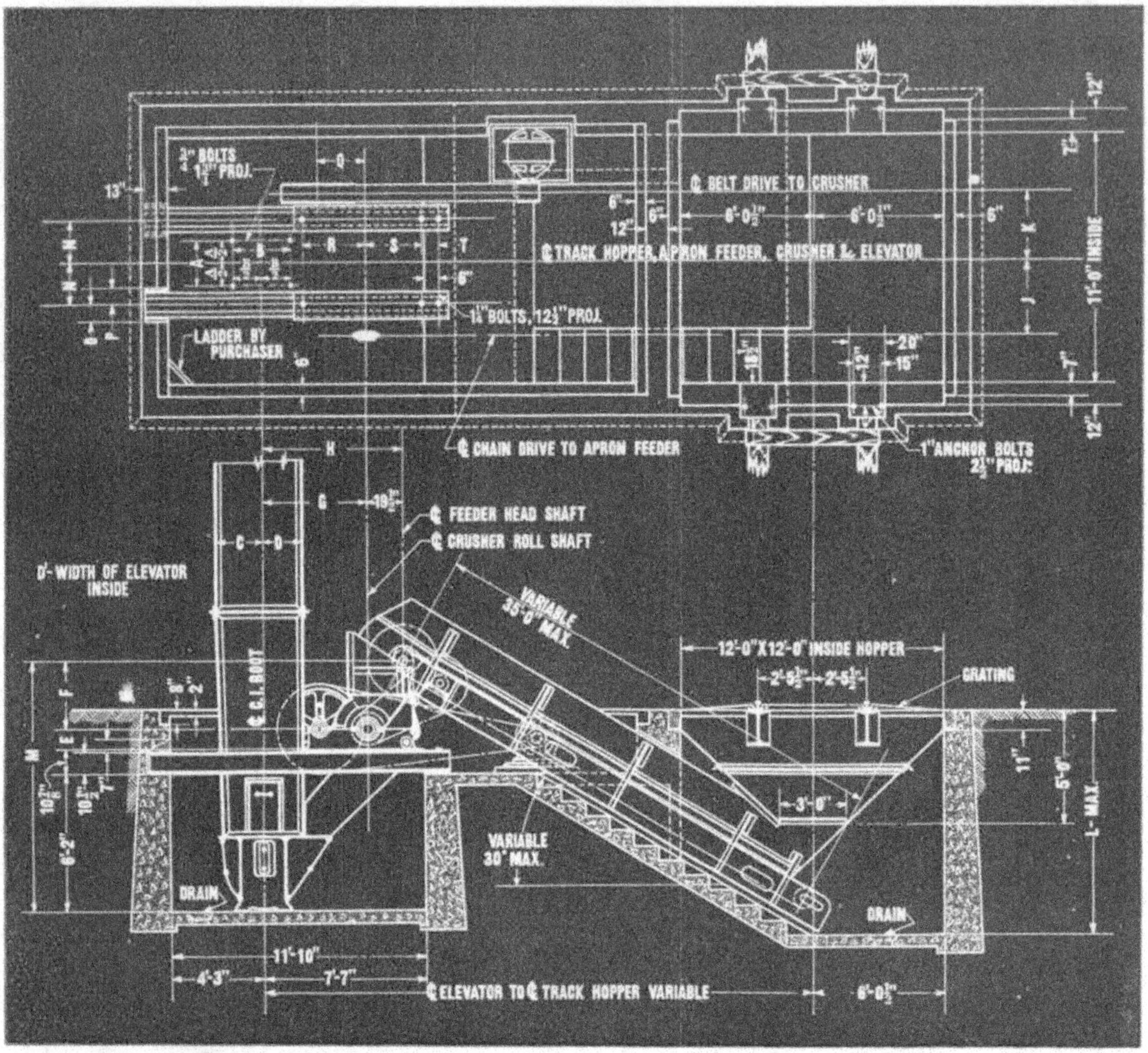

General arrangement of a Jeffrey 12 x 12 Steel Track Hopper with Apron Feeder to handle Coal through Single Roll Crusher to Bucket Elevator.

ELEVATORS							
Capacity Tons per Hour	No.	Centers	A	B	C	D	D1
23.2	115 144	0′-40′ 41′-80′	19½″ 20½″	20″ 20″	2′-1′ 2′-5″	19″ 19″	15″ 16″
25	119 149	0′-40′ 41′-80′	21½″ 22½″	20″ 20″	2′-1″ 2′-5″	19″ 19″	17″ 18″
36	122 152	0′-40′ 41′-80′	2′-0″ 2′-1″	2′-0″ 2′-0″	2′-0″ 2′-2″	22″ 22″	19″ 20″
50	Cont. Bucket	0′-40′ 41′-80′	2′-0″ 2′-1″	2′-0″ 2′-0″	2′-0″ 2′-2″	22″ 22″	19″ 20″

Where local conditions will not permit of installing a standard 12 x 12 Hopper, an 8 x 8 or 10 x 10 Hopper can be furnished.

Full information on request.

‡Grating omitted when maximum size of piece does not exceed tabulated dimensions.

Purchaser to determine thickness of pit walls to suit local soil conditions.

For detailed information on Bucket Elevators, see pages 64 to 69.

For detailed information on Single Roll Crusher, see pages 185 to 194.

Feeder Using Chain No.	Size Crusher	Motor Req'd HP	Motor Req'd Speed	E	F	G	H	J	K	L	M	N	O	P	Q	R	S	T	Mesh of Grating ‡
126 C 156 C or 951 S T R	24x24	25	860	10½″	3′-0″	4′-8⅛″	6′-3⅝″	2′-9″	2′- 4¾″	10′-0″	10′-11⅜″	17⅞″	8″	6″	21″	2′- 5½″	2′-2″	14⅞″	14″
	30x30	35	860	12″	3′-1″	4′-9½″	6′-5″	3′-3″	2′-11⅜″	10′-0″	11′- 1⅞″	22⅛″	9″	9″	2′-1½″	2′-10¾″	2′-6¾″	8¾″	20″
809 S T R	24x24	25	860	10½″	3′-3″	4′-8⅛″	6′-3⅝″	2′-9″	2′- 4¾″	10′-6″	11′- 2⅜″	17⅞″	8″	6″	21″	2′- 5½″	2 -2″	14⅞″	14″
	30x30	35	860	12″	3′-4″	4′-9½″	6′-5″	3′-3″	2′-11⅜″	10′-6″	11′- 4⅞″	22⅛″	9″	9″	2′-1½″	2′-10¾″	2′-6¾″	8¾″	20″

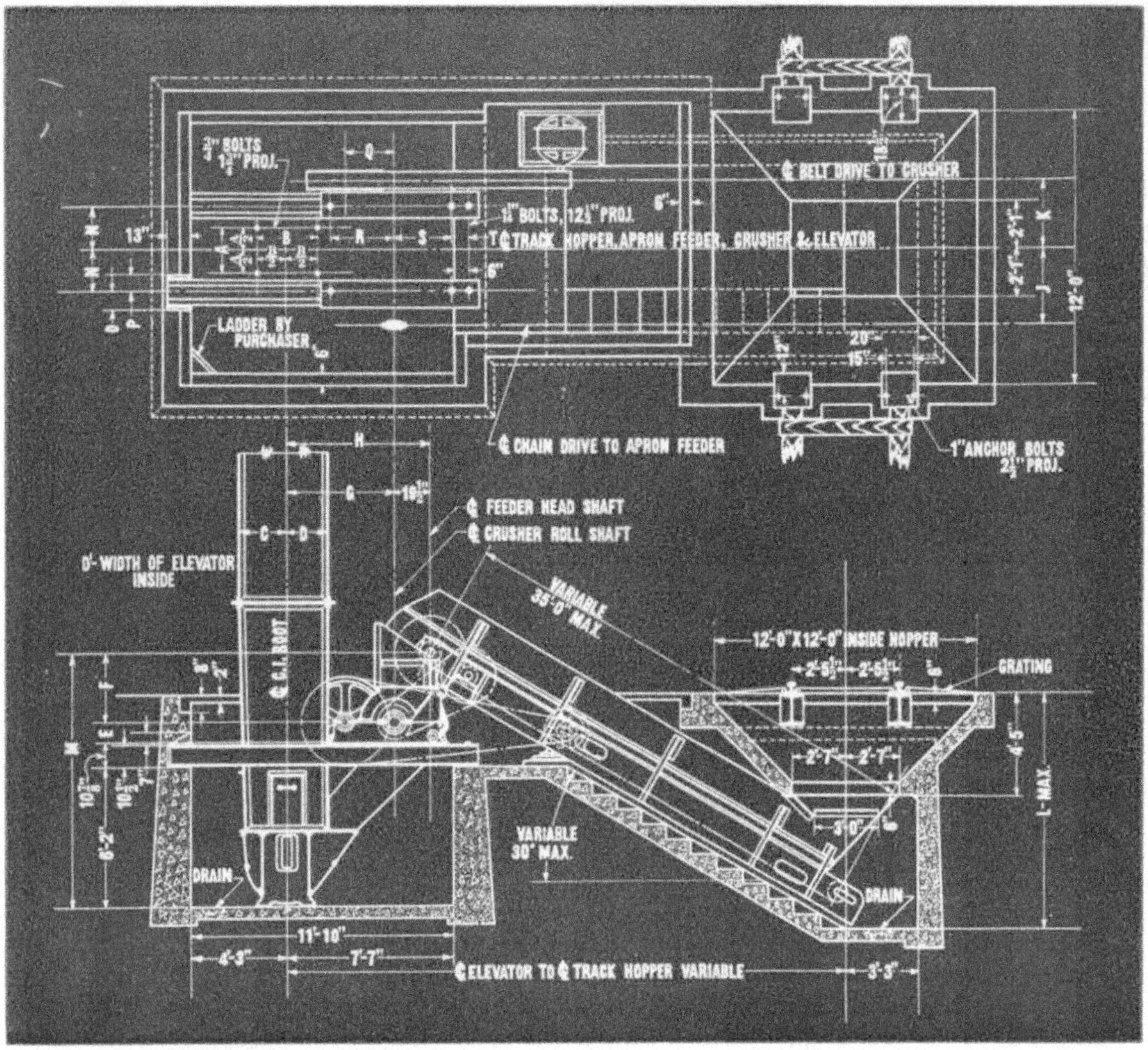

General arrangement of a Jeffrey 12 x 12 Concrete Track Hopper with Apron Feeder to handle Coal through Single Roll Crusher to Bucket Elevator.

ELEVATORS

Capacity Tons per Hour	No.	Centers	A	B	C	D	DI
23.2	115 144	0'-40' 41'-80'	19½" 20½"	20" 20"	2'-1" 2'-5"	19" 19"	15" 16"
25	119 149	0'-40' 41'-80'	21½" 22½"	20" 20"	2'-1" 2'-5"	19" 19"	17" 18"
36	122 152	0'-40' 41'-80'	2'-0" 2'-1"	2'-0" 2'-0"	2'-0" 2'-2"	22" 22"	19" 20"
50	Cont. Bucket	0'-40' 41'-80'	2'-0" 2'-1"	2'-0" 2'-0"	2'-0" 2'-2"	22" 22"	19" 20"

Where local conditions will not permit of installing a standard 12 x 12 Hopper, an 8 x 8 or 10 x 10 Hopper can be furnished.

Full information on request.

‡Grating omitted when maximum size of piece does not exceed tabulated dimensions.

Purchaser to determine thickness of pit walls to suit local soil conditions.

For detailed information on Bucket Elevators, see pages 64 to 69.

For detailed information on Single Roll Crusher, see pages 185 to 194.

Feeder Using Chain No.	Size Crusher	Motor Req'd HP	Motor Req'd Speed	E	F	G	H	J	K	L	M	N	O	P	Q	R	S	T	‡Mesh of Grating
126 (156 (or 951 S T R	24x24	25	860	10½"	3'-0"	4'-8⅜"	6'-3⅝"	2'-9"	2'- 4¾"	10'- 5"	10'-11⅜"	17⅞"	8"	6"	21"	2'- 5½"	2'-2"	14⅞"	14"
	30x30	35	860	12"	3'-1"	4'-9½"	6'-5"	3'-3"	2'-11⅜"	10'- 5"	11'- 1⅞"	22⅛"	9"	9"	2'-1½"	2'-10¼"	2'-6¾"	8¾"	20"
809 S T R	24x24	25	860	10½"	3'-3"	4'-8⅜"	6'-3⅝"	2'-9"	2'- 4¾"	10'-11"	11'- 2⅜"	17⅞"	8"	6"	21"	2'- 5½"	2'-2"	14⅞"	14"
	30x30	35	860	12"	3'-4"	4'-9½"	6'-5"	3'-3"	2'-11⅜"	10'-11"	11'- 4⅞"	22⅛"	9"	9"	2'-1½"	2'-10¼"	2'-6¾"	8¾"	20"

Jeffrey Rack and Pinion Bin Valves as installed in a large Power House.

Jeffrey Rack and Pinion Bin Valve

Valve Plate operates on rollers as shown in line drawing below.

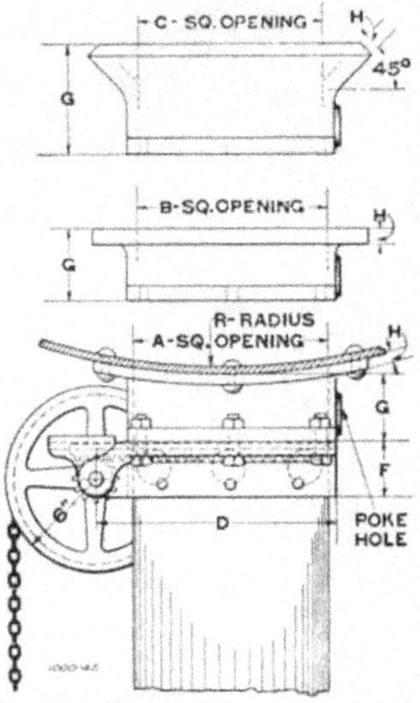

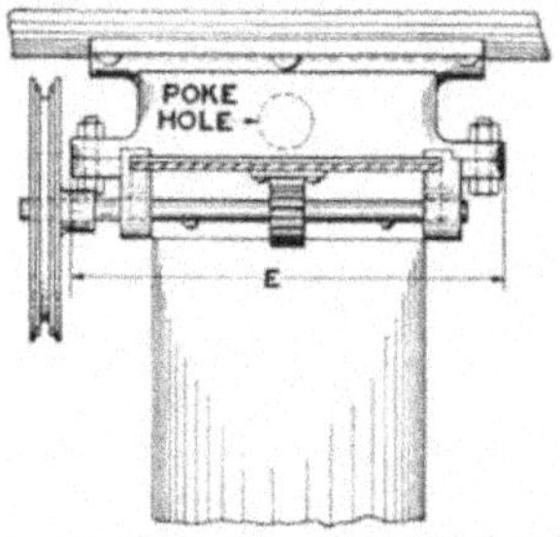

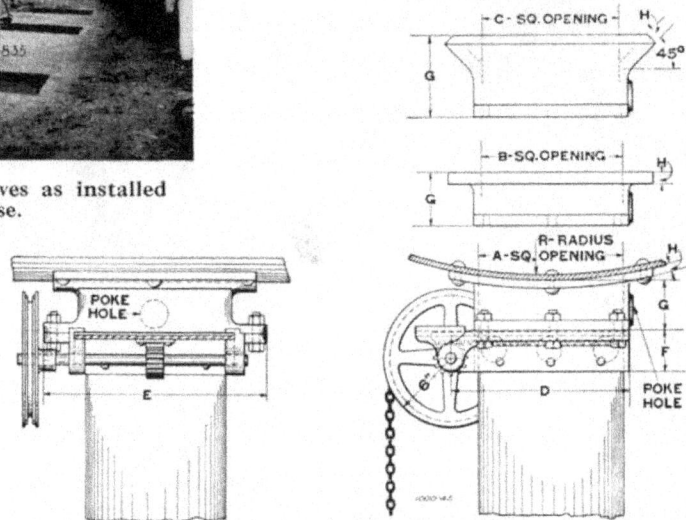

RACK and Pinion Bin Valves are extensively used in connection with Dump Hoppers, Storage Bins, etc. They are rugged and substantial in every detail and by means of the great leverage secured through the hand or chain wheel in connection with the gear pinion and rack under the valve plate a large closing pressure may be readily secured.

Valves are furnished with 12" hand wheels or 12" pocket sheaves as ordered. The operating chain for pocket sheaves is extra. Where the valve is to be operated at a distance, just sufficient chain may be secured to operate the valve, the free ends of the chain being connected to extension wires or ropes.

In practice it has been found that bolt holes in bin should be punched in field to match valves.

Dimensions of Rack and Pinion Bin Valve

Style	Item No.	List Price* Valve Plate resting upon rollers	Approx. Weight Lbs.	Dimensions in Inches									Pattern No. of Nozzle at Bin
				A	B	C	D	E	F	G	H	R	
Curved Flange Nozzle	1	See Price List Bulletin	350	14			17⅝	22¼	5	4⅝	⅞	54	24361
	2		350	14			17⅝	22¼	5	4⅝	⅞	45	19883
	3		350	14			17⅝	22¼	5	4⅝	⅞	36	17602
	4		350	14			17⅝	22¼	5	4⅝	⅞	18	18320
	5		350	14			17⅝	22¼	5	4⅝	⅞	85	61468
	6		350	14			17⅝	22¼	5	4⅝	⅞	60	60419
	7		350	14			17⅝	22¼	5	4⅝	⅞	54 5/16	8140
	8		500	20			23¾	28¾	5	6	1⅛	45	19217
Flat Flange Nozzle	9		350		14		17⅝	22¼	5	4⅝	⅞		9477
	10		350		14		17⅝	22¼	5	6⅝	⅞		195[illegible]8
	11		500		20		23¾	28¾	5	7	1⅛		195[illegible]7
45° Bevel Flange Nozzle	12		350			14	17⅝	22¼	5	6¼	⅞		18883

*In ordering give Page and Item Number.

The illustration at left shows a Jeffrey 20″ x 20″ Clam Shell Valve used to load directly from an overhead storage bin of 2000 cubic feet capacity, into railroad cars.

JEFFREY Clam Shell Valves are of the same rugged construction and apply to the same service as the Rack and Pinion Valves on preceding page. They are also extensively used in foundries and other industries where the valve is constantly in service and where a quick opening and closing of the clams is required. Operating chain and brackets are extra and are not furnished unless requested.

In practice it has been found best that bolt holes in bin should be punched in field to match valves.

We can furnish 24″ valves with steel body variable to 48″ maximum width. Additional Information on Application.

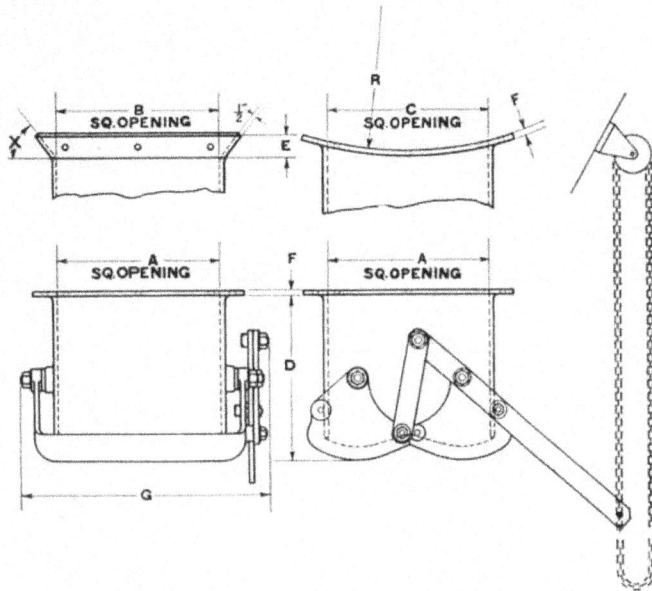

General Dimensions
For Prices, See Price List Bulletin

Style of Top	A	B	C	D	E	F	G	R	X	Approx. Weight Lbs.	Pattern No. of Body
Flat Flange	16″			15¾″		¾″	2′- 1⅝″			361	62100
	20″			19¾″		¾″	2′- 7″			547	62087
	24″			22⅞″		¾″	2′-10¾″			588	62351
Bevel Flange		16″		15¾″	2½″		2′- 1⅝″		52°	345	62112
		20″		19¾″	2½″		2′- 7″		52°	537	62355
		24″		22⅞″	2½″		2′-10¾″		52°	577	62353
Curved Flange			16″	15¾″		¾″	2′- 1⅝″	4′-0″		380	62094
			20″	19¾″		¾″	2′-7″	4′-0″		558	62354
			24″	22⅞″		¾″	2′-10¾″	4′-0″		608	62352

Slide Valves for General Service ═══ JEFFREY═

Applicable to all kinds of Elevators, Conveyors, Bins and Hoppers
Dimensions and Prices on Application.

An application of Jeffrey Plain Slide Valve in connection with Spiral Conveyor.

Jeffrey Plain Rack and Pinion Slide Valve, operating in connection with Scraper Conveyor.

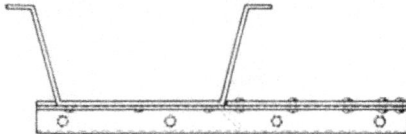

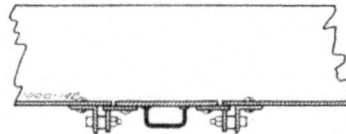

Figure 1.

Figure 1.
Plain Slide Valve, may be operated by links with lever in place of hand grip.

Figure 2.
Roller Bearing Slide Valve, similar to Fig. 1 used especially under sticky, corrosive or wet freezing conditions.

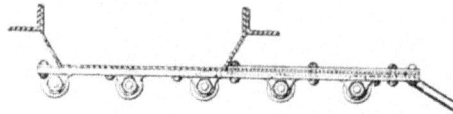

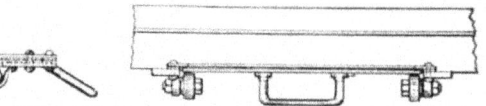

Figure 2.

Figure 3.

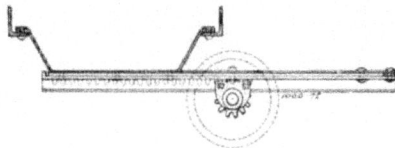

Figure 3.
Plain Rack and Pinion Slide Valve, operated by hand or chain wheel. Can be equipped with rollers same as Fig. 2.

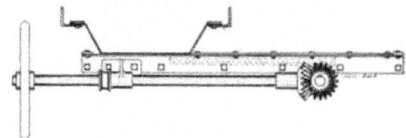

Figure 4.
Bevel Gear Operated Rack and Pinion Valve with horizontal operating shaft. The operating shaft may be placed vertical if desired.

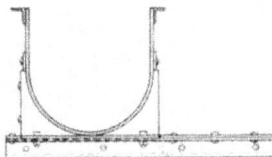

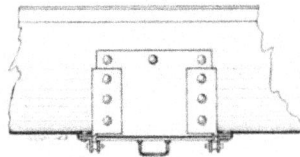

Figure 5.

Figure 5.
Plain Slide Valve as applied cross-wise to Spiral Conveyor trough. Length of opening in trough for free discharge should be at least 1¼ times the diameter of the spiral.

Figure 6.
Plain Slide Valve rolled to fit length-wise of Spiral Conveyor trough. Valve to be placed to pull against the flow of material in trough.

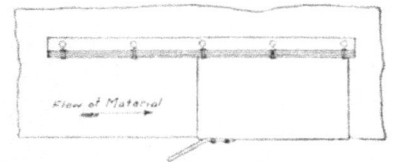

Figure 6.

For other Slide Valves, see page 96.

90

Scraper
Conveyors

Section

6

A Jeffrey Scraper Conveyor consisting of steel scrapers with malleable iron wearing blocks, mounted on a single strand of Flat and Round Steel Link Chain, is here shown distributing coal over bunkers.

Here Malleable Iron Scrapers are mounted upon a single strand of Jeffrey Detachable Chain. This same type of scraper is also furnished mounted on steel Vulcan Chain.

Scraper Conveyors as shown above consisting of all steel scrapers mounted between two strands of high class Steel Thimble Roller Chain are used where large capacities or severe service are encountered.

Jeffrey Scraper Conveyor made up with Steel Scrapers mounted upon rollers and propelled by a Single Strand of all steel vulcan type chain, has always proved its worth in a most satisfactory service wherever it has been installed for the handling of coal in boiler houses.

Scraper Conveyor with double strand of Malleable Roller chain installed in a Power House for the handling of coal from track hopper to bunkers.

Double Strand Vulcan Chain Scraper Conveyor fed by Bucket Elevator handling coal in a large Steel Plant.

Some Important Points to assist you in Selecting a Jeffrey Standard Scraper Conveyor

THE Scraper Conveyor is made of both single and double strands of chains, according to the size of scraper. The single strand being limited to 18-inch maximum length of scrapers, while the double strands, although limited to no particular length of scrapers, seldom are called upon to take lengths greater than 36 inches.

Simple construction a desirable feature

Single Strand Conveyors have the chain bolted to the top of the scrapers at their centers. Scrapers for this type are ordinarily made of either malleable iron with self-contained wearing surfaces or of steel plate with wearing blocks or roller attachments at their ends.

The Double Strand Conveyors have the chains bolted to the ends of the scrapers by means of extension attachments of the chains, or as in the Vulcan type of Chain, by means of long steel bars bolted to the scrapers at the top and bent out at the ends to form the side bars of the chains.

Have a Place in nearly every Industry

Both the Single and Double Strand Conveyors are designed to handle all kinds of loose products of the farm, manufacturing and mining industries, with their widest application being unquestionably given to the handling of coal and similar semi-abrasive loose materials. Note the wide range of application of scraper conveyors as illustrated in the preceding and the following pages.

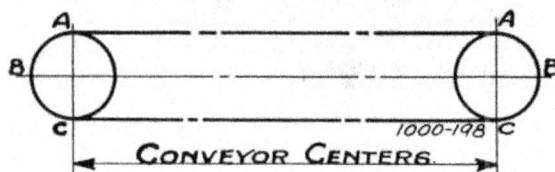

Wide Range of Service from One Conveyor

Scraper Conveyors may be installed on the horizontal, on an incline, or as a combination in one conveyor of both horizontal and incline, or of two inclines, with the joining connection between the horizontal and incline or the two inclines being made in a curve of large radius. The angle of incline of a Scraper Conveyor either singly or in combination should not exceed 45 degrees and preferably not over 30 degrees to 35 degrees to the horizontal. The Standard Scraper Conveyors, as listed in the tables, are divided into four groups, based upon their length or centers.

For the Single Strand Conveyors, three types of chain, Vulcan, Detachable and Steel Link have been found to be most suitable and so have become the Jeffrey Standard. On the Double Strand Conveyors the same types of chain are used as on the Single Strand, with the addition of the Malleable Roller and Steel Thimble Roller types, this latter type having been found especially efficient in the handling of the larger capacities.

In addition to the foregoing Single and Double Strand Scrapers is the Drag Chain, which is essentially a Single Strand Scraper within the chain itself, the chain being sufficiently wide to form a scraper.

Total Shipping Weights easily figured from Tables

The "Weight of Terminals" given in the tables includes shafts, bearings, collars, sprockets and gears, with chain and flights half way around the sprockets as shown by the arcs A B C in the sketch at the left.

Index to Conveyors Built on Wood Supports

Average Size Material to be Handled	Maximum Size Pieces	Capacity in Tons* per Hour Horizontal	Size of Scraper	0 to 50 ft. Centers		51 to 100 ft. Centers		101 to 150 ft. Centers		151 to 200 ft. Centers	
				Conveyor No.	Page No.	Conveyor No.	Page No.	Conveyor No.	Page No.	Conveyor No.	Page No.
Single Strand Chain with Malleable Iron Scrapers											
1½	3	42	10x 5	2924	98	2929	98	2934	98		
2	4	50	12x 5	2925	98	2930	98	2935	98		
2	4	50	12x 5	2926	98	2931	98				
3	5	63	15x 5	2927	98	2932	98	2936	98		
3	5	63	15x 5	2928	98	2933	98	2937	98		
Single Strand Chain with Roller Attachments											
1¾	3½	48	15x 7	2938	106	2942	106	2946	106	2950	106
1¾	3½	48	15x 7	2939	106	2943	106	2947	106	2951	106
3	5	70	18x 8	2940	106	2944	106	2948	106	2952	106
3	5	70	18x 8	2941	106	2945	106	2949	106		
Single Strand Chain with Wearing Blocks											
1¾	3½	48	15x 7	3300	102	3304	102	3308	102	3312	102
1¾	3½	48	15x 7	3301	102	3305	102	3309	102	3313	102
3	5	70	18x 8	3302	102	3306	102	3310	102	3314	102
3	5	70	18x 8	3303	102	3307	102	3311	102		
Double Strand Malleable Roller Chain											
6	9	60	18x 6	2953	110	2956	110	2959	110	2962	110
6	9	60	18x 6	2954	110	2957	110	2960	110	2963	110
8	12	112	24x 8	2955	110	2958	110	2961	110	2964	110
Double Strand Vulcan Chain											
6	9	60	18x 6	2965	114	2968	114	2971	114	2973	114
8	12	112	24x 8	2966	114	2969	114	2972	114		
Double Strand Steel Thimble Roller Chain											
8	12	92	24x 8	2982	118	2985	118	2988	118	2991	118
10	14	167	30x10	2983	118	2986	118	2989	118	2992	118
12	16	238	36x12	2984	118	2987	118	2990	118	2993	118

Index to Conveyors Built on Steel Supports

Average Size Material to be Handled	Maximum Size Pieces	Capacity in Tons* per Hour Horizontal	Size of Scraper	0 to 50 ft. Centers		51 to 100 ft. Centers		101 to 150 ft. Centers		151 to 200 ft. Centers	
				Conveyor No.	Page No.	Conveyor No.	Page No.	Conveyor No.	Page No.	Conveyor No.	Page No.
Single Strand Chain with Malleable Iron Scrapers											
1½	3	42	10x 5								
1½	3	42	10x 5	2994	100	2999	100	3004	100		
2	4	50	12x 5	2995	100	3000	100	3005	100		
2	4	50	12x 5	2996	100	3001	100	3006	100		
3	5	63	15x 5	2997	100	3002	100	3007	100		
3	5	63	15x 5	2998	100	3003	100				
Single Strand Chain with Roller Attachments											
1¾	3½	48	15x 7	3008	108	3012	108	3139	108	3143	108
1¾	3½	48	15x 7	3009	108	3013	108	3140	108	3144	108
3	5	70	18x 8	3010	108	3137	108	3141	108	3145	108
3	5	70	18x 8	3011	108	3138	108	3142	108		
Single Strand Chain with Wearing Blocks											
1¾	3½	48	15x 7	3315	104	3319	104	3323	104	3327	104
1¾	3½	48	15x 7	3316	104	3320	104	3324	104	3328	104
3	5	70	18x 8	3317	104	3321	104	3325	104	3329	104
3	5	70	18x 8	3318	104	3322	104	3326	104		
Double Strand Malleable Roller Chain											
6	9	60	18x 6	3146	112	3149	112	3152	112	3155	112
6	9	60	18x 6	3147	112	3150	112	3153	112	3156	112
8	12	112	24x 8	3148	112	3151	112	3154	112	3157	112
Double Strand Vulcan Chain											
6	9	60	18x 6	3158	116	3161	116	3164	116	3166	116
8	12	112	24x 8	3159	116	3162	116	3165	116		
Double Strand Steel Thimble Roller Chain											
8	12	92	24x 8	3175	120	3178	120	3181	120	3184	120
10	14	167	30x10	3176	120	3179	120	3182	120	3185	120
12	16	238	36x10	3177	120	3180	120	3183	120	3186	120

* 50 lbs. per. cu. ft.

How to Select a Jeffrey Scraper Conveyor to Meet Your Conditions

JEFFREY Conveyors are built under four different groups of lengths or "centers." The 1st group covers all conveyors up to 50 ft. in length, the 2nd from 51 to 100 ft., the 3rd from 101 to 150 ft. and the last group covering lengths from 151 to 200 ft.

Example: In selecting a conveyor, with the "Average Size of Material" 3 inch pieces, the "Maximum Size" 5 inch pieces, the "Capacity" requirement 70 tons per hour, and the length 125 ft. Under the first and second columns you will see that six groups of Conveyors will handle these sizes, but coming to the 3rd column you find that only 4 of these groups of Conveyors have a capacity of 70 tons per hour. Your Conveyor must be 125 ft. long so you go over the column "101 to 150 ft. Centers" and we find Conveyors No. 2948 and 2949, details of which are given in table on page 106, also No. 3310 and 3311 given in table on page 102, will take care of your requirements. The first mentioned are fitted with roller attachments, the latter with wearing blocks as shown in table on page 95 for Wood Supports.

Suppose your "Average" was 1¾ inch pieces, your "Maximum" 9 inch pieces and your "Capacity" was "40 tons per hour," you would have a choice of three different Conveyors, using double strands of Chain, Nos. 2959 and 2960 in table on page 110 or No. 2971 given in table on page 114.

By consulting the tables it will be noted that the capacities of the Conveyors No. 2959 and 2960, which are capable of handling these maximum size pieces, are far in excess of the requirements. In such cases it is the size pieces rather than the capacity which governs the selection of Conveyor. Under these conditions the speed of the conveyor may be reduced in direct proportion, thereby materially increasing the life of the Conveyor, and reducing the horse power required proportionately to the reduction of speed.

Elements Affecting Capacities and Horsepower

WHERE a capacity of scraper on an incline is listed as "zero," the safe working strength of the chain has been exceeded for the "Centers" of Conveyor at the head of the Table, and heavier equipment must be selected.

When the capacity of a combination Horizontal and Incline Conveyor is desired be sure to use the capacity rating given in the Tables for the Incline.

The "Horsepower at the Countershaft" should ordinarily be increased 10% for each speed reduction of gears or belting between this Countershaft and the Motor or Engine. In this connection it is best to use not less than a 3 H. P. Motor for motors figuring less than 3 Horsepower.

Valves for Jeffrey Scraper Conveyors

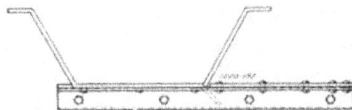

Fig. 1. Plain Hand Slide Valve.

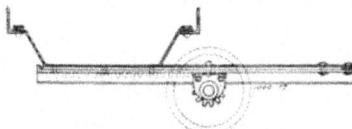

Fig. 2. Rack and Pinion Slide Valve.

THE Plain Slide Type of Valve can be used on any Standard Scraper but is ordinarily limited to Conveyors having Scrapers not over 24 inches long. Valves Figures 2, 3, and 4, also may be furnished with any Standard Scraper Conveyor, but are ordinarily not used on Con-

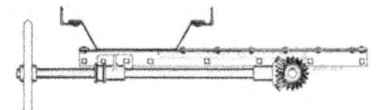

Fig. 3. Bevel Gear operated Rack and Pinion Valve, horizontal operating shaft.

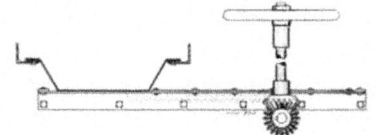

Fig. 4. Bevel Gear operated Rack and Pinion Valve, vertical operating shaft.

veyors having Scrapers smaller than 15 inches wide by 7 inches deep.

In this whole matter of valves, it is to be noted that when a choice of valves is not expressed by the Purchaser, in his choice of a Scraper Conveyor, Jeffrey Engineers will select that type of valve and spacing, which in their judgment is best fitted to the Purchaser's statements or sketches of his requirements.

Selecting Conveyors of Irregular Contour from Tables

THE greater number of all Scrapers are installed on the horizontal, but many of the most profitable applications of the Scraper Conveyor are combinations of both the horizontal and incline in as much as such combinations usually take the place of two or more separate units and at a much less initial cost

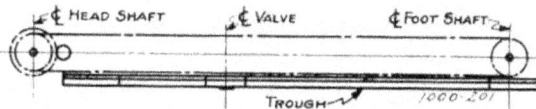

Fig. 5. Horizontal Conveyor for along ground or over storage bins.

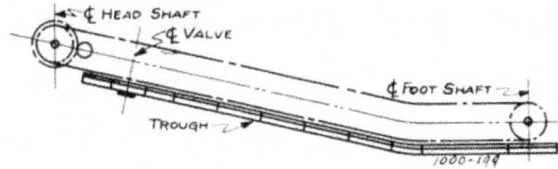

Fig. 7. Combination Conveyor, up slope from receiving hopper and over bins.

and upkeep, see illustrations upon pages 92, and 93.

Assume our problem is to handle 60 Tons per hour of bituminous coal of 6-inch "Average Size Material" and 12-inch "Maximum Size Pieces," 40 feet up a 30 degree incline and 96 feet horizontal over a bin storage, and to be erected upon "Wood Supports."

Consulting "Index to Conveyors," page 95, we find that the 12-inch "Maximum Size Pieces" control the choice of the smallest Conveyors, which may be used and Tables for which are given on pages 110, 114 and 118. Examining these pages we find 44 tons to be the largest capacity up 30 degree slope on pages 110

and 114, with 93 tons the limit on page 118.

The Scraper Conveyor required is therefore on page 118. The next step is to find those centers of horizontal conveyors as given at the top of the Tables which correspond to the combination of 96 feet horizontal and 40 feet incline. To do this we reduce

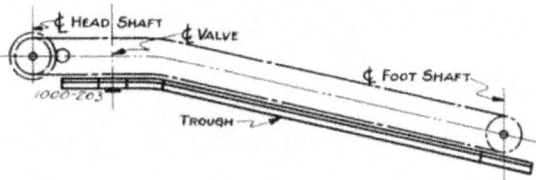

Fig. 6. Combination Conveyor, horizontal receiving section and slope to pile.

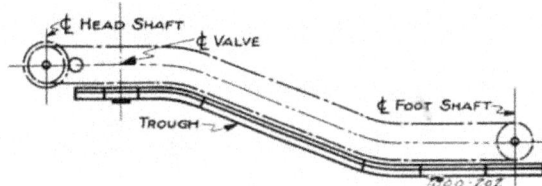

Fig. 8. Combination Conveyor, receiving section, up slope and over storage.

complicated figuring to the following simple formula: $(.75H) + (B \times J) = L$, where H is the horizontal length, J the inclined length, and L equivalent length of Conveyor for strength, while B is 1.5 for Scraper Conveyors having chains sliding, 1.8 for Malleable Roller Chains and 2.0 for Steel Thimble Roller Chains or Single Strand Conveyors with Scrapers fitted with rollers.

Substituting in this formula for page 118, we have $(.75$ of $96) + (2.0$ of $40) = 152$ feet centers, that is to say Conveyor No. 2992 listed for 65 Tons on 30 degree incline and ordered 136 ft. centers will completely fill our requirements.

Large Curves Reduce Wear and Save Power

AT this point it is to be noted that it is only Scraper Conveyors of the double strand type which can be used to the best advantage in combination of horizontal and inclines. The runways which support the chains on either side of such Scrapers when used in conjunction with hold-down guides serve to keep the scrapers down into place as the scrapers pass around curves from horizontal to incline or vice versa. Such curves in good practice should ordinarily have a radius of curvature to the centers of the chains of not less than one foot for each inch of chain pitch. That is to

say six foot radius for a six inch pitch chain, eight foot for eight inches, twelve foot for twelve inches, etc. As it is quite obvious, however, that the larger the curve radius the less wear there will be on the chains and guides and also the less power there will be consumed due to reduced friction, it should always be the purpose to make curves as large as possible to average maximum inclines of 30 to 40 degrees, being limited only by local conditions or by the placing of valves. Valves should always be located in straight rather than curved sections of a scraper trough.

Specifications of Jeffrey Standard Scraper Conveyors Using Single Strand Detachable and Vulcan Chain with Malleable Scrapers

Wood Supports—*For Steel Supports see page 100.*

Length of Conveyor	0 to 50 ft. Centers					51 to 100 ft. Centers					101 to 150 ft. Centers			
No. of Conveyor	2924	2925	2926	2927	2928	2929	2930	2931	2932	2933	2934	2935	2936	2937
Size of Material—Inches														
Average size of Material to be handled	$1\frac{1}{2}$	2	2	3	3	$1\frac{1}{2}$	2	2	3	3	$1\frac{1}{2}$	2	3	3
Maximum size; not to exceed 10% of whole	3	4	4	5	5	3	4	4	5	5	3	4	5	5
Capacity—In tons per hr.														
Horizontal	42	50	50	63	63	42	50	50	63	63	42	50	63	63
15° Incline	23	27	27	33	33	23	27	27	33	33	0	27	33	0
30° Incline	16	20	20	24	24	16	20	20	24	24	0	20	0	0
45° Incline	14	17	17	20	20	14	17	17	20	20	0	0	0	0
Size Scraper—Inches														
Length	10	12	12	15	15	10	12	12	15	15	10	12	15	15
Depth	5	5	5	5	5	5	5	5	5	5	5	5	5	5
Mall. Iron Pattern No.	28026	27898	27898	28656	28656	28026	27898	27898	28656	28656	28026	27898	28656	28656
Spacing—Inches	26	26	24	24.56	24	26	26	24	24.56	24	26	24	24.56	24
Chain														
Number and Style	88J	88J	526V	103J	526V	88J	88J	526V	103J	526V	88J	526V	103J	526V
Pitch—Inches	2.6	2.6	6.0	3.07	6.0	2.6	2.6	6.0	3.07	6.0	2.6	6.0	3.07	6.0
Attachments	F-2	F-2	A½	F-2	A½	F-2	F-2	A½	F-2	A½	F-2	A½	F-2	A½
Working Strength—Lbs.	960	960	1640	1600	1640	960	960	1640	1600	1640	960	1640	1600	1640
H. P. At Countershaft*	1.3	1.5	1.8	1.9	2.1	2.6	3.0	3.6	3.8	4.2	3.9	5.3	5.8	6.2
Head Shaft														
Diameter—Inches	$1\frac{15}{16}$	$1\frac{15}{16}$	$1\frac{15}{16}$	$1\frac{15}{16}$	$2\frac{7}{16}$	$2\frac{7}{16}$	$2\frac{7}{16}$	$2\frac{7}{16}$	$2\frac{7}{16}$	$2\frac{15}{16}$	$2\frac{7}{16}$	$2\frac{15}{16}$	$2\frac{15}{16}$	$2\frac{15}{16}$
Rev. per Min.	$16\frac{2}{3}$	$16\frac{2}{3}$	$16\frac{2}{3}$	$16\frac{2}{3}$	$16\frac{2}{3}$	$16\frac{2}{3}$	$16\frac{2}{3}$	$16\frac{2}{3}$	$16\frac{2}{3}$	$16\frac{2}{3}$	$16\frac{2}{3}$	$16\frac{2}{3}$	$16\frac{2}{3}$	$16\frac{2}{3}$
Size Sprocket—Inches	23	23	23½	23½	23½	23	23	23½	23½	23½	23	23½	23½	23½
Gear Diameter—Inches	29.83	29.83	29.83	29.83	29.83	29.83	29.83	29.83	29.83	29.83	29.83	29.83	29.83	29.83
Gear Pitch—Inches	1¼	1¼	1¼	1¼	1¼	1¼	1¼	1¼	1¼	1¼	1¼	1¼	1¼	1¼
Gear Face—Inches	3	3	3	3	3	3	3	3	3	3	3	3	3	3
Countershaft														
Diameter—Inches	$1\frac{7}{16}$	$1\frac{7}{16}$	$1\frac{7}{16}$	$1\frac{7}{16}$	$1\frac{15}{16}$	$1\frac{15}{16}$	$1\frac{15}{16}$	$1\frac{15}{16}$	$1\frac{15}{16}$	$2\frac{7}{16}$	$1\frac{15}{16}$	$2\frac{7}{16}$	$2\frac{7}{16}$	$2\frac{7}{16}$
Rev. per Min.	83	83	83	83	83	83	83	83	83	83	83	83	83	83
Pinion Diameter—Inches	6.01	6.01	6.01	6.01	6.01	6.01	6.01	6.01	6.01	6.01	6.01	6.01	6.01	6.01
Pinion Face—Inches	3¼	3¼	3¼	3¼	3¼	3¼	3¼	3¼	3¼	3¼	3¼	3¼	3¼	3¼
Foot Shaft														
Diameter—Inches	$1\frac{7}{16}$	$1\frac{7}{16}$	$1\frac{7}{16}$	$1\frac{7}{16}$	$1\frac{7}{16}$	$1\frac{7}{16}$	$1\frac{7}{16}$	$1\frac{7}{16}$	$1\frac{7}{16}$	$1\frac{15}{16}$	$1\frac{7}{16}$	$1\frac{15}{16}$	$1\frac{15}{16}$	$1\frac{15}{16}$
Size Sprocket—Inches	23	23	23½	23½	23½	23	23	23½	23½	23½	23	23½	23½	23½
Trough														
Thickness or Gauge	10	10	10	10	$\frac{3}{16}$	10	10	10	10	$\frac{3}{16}$	10	10	10	$\frac{3}{16}$
Approx. Shipping Wgt.—Lbs.														
Terminals, Complete	630	630	680	670	770	730	730	770	760	950	730	930	920	950
Chain and Flights Per Ft. Ctrs.	10	9	16½	13	17	10	9	16½	13	17	10	16½	13	17
Trough and Bar Trackage Per Ft. Ctrs.	11	12	12	13½	17	11	12	12	13½	17	11	12	13½	17

*For Maximum Centers for all inclinations and corresponding capacities.

"Weight of Terminals" given in table includes shafts, bearings, collars, sprockets and gears, with chain and flights half way around the sprockets.

Capacities are figured for 50 lb. material, troughs 80% level full, and uniformly loaded throughout the time period.

For Erection Dimensions of above conveyors see page 99.

General Dimensions of Jeffrey Standard Scraper Conveyors

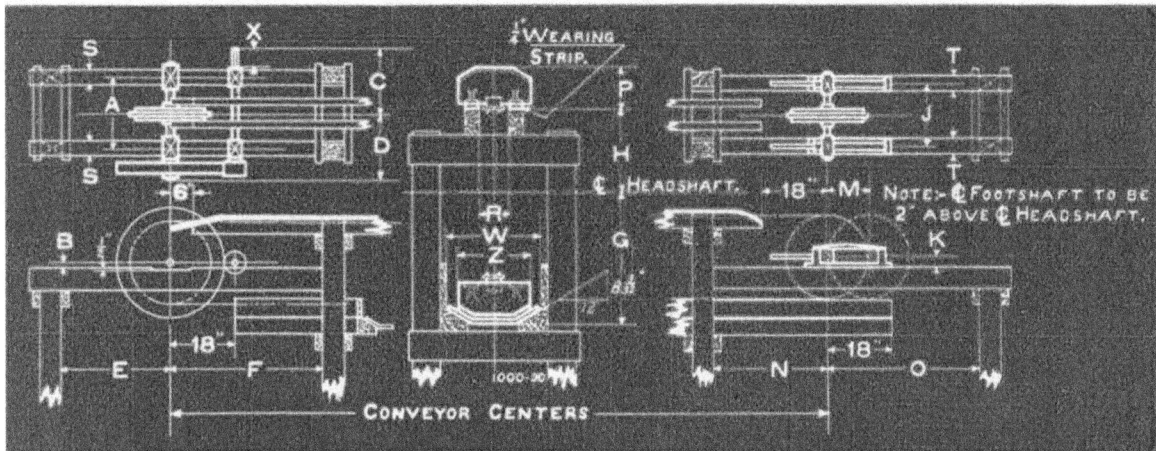

Using Single Strand Detachable and Vulcan Chain with Malleable Scrapers

Wood Supports—*For Steel Supports see page 101.* Dimensions in Inches.

Conveyor No.	A	B	C	D	E	F	G	H	J	K	M	N	O	P	R	S	T	W	X	Z
2924	19	$1\frac{3}{16}$	$17\frac{3}{4}$	$16\frac{3}{4}$	30	42	$17\frac{1}{8}$	$11\frac{3}{4}$	17	$2\frac{1}{4}$	$11\frac{3}{4}$	30	42	$5\frac{1}{4}$	4	4	4	13	6	10
2925	21	$1\frac{3}{16}$	$18\frac{3}{4}$	$17\frac{3}{4}$	30	42	$17\frac{1}{8}$	$11\frac{3}{4}$	19	$2\frac{1}{4}$	$11\frac{3}{4}$	30	42	$5\frac{1}{4}$	4	4	4	15	6	12
2926	21	$1\frac{3}{16}$	$18\frac{3}{4}$	$17\frac{3}{4}$	30	42	$17\frac{5}{8}$	$12\frac{1}{4}$	19	$2\frac{1}{4}$	$11\frac{3}{4}$	30	42	$5\frac{1}{4}$	4	4	4	15	6	12
2927	24	$1\frac{3}{16}$	$20\frac{1}{4}$	$19\frac{3}{4}$	30	42	$17\frac{1}{2}$	$12\frac{1}{8}$	22	$2\frac{1}{4}$	$11\frac{3}{4}$	30	42	$5\frac{1}{4}$	4	4	4	18	6	15
2928	24	$2\frac{5}{8}$	21	$20\frac{3}{4}$	30	42	$17\frac{11}{16}$	$12\frac{1}{4}$	22	$2\frac{3}{4}$	$11\frac{3}{4}$	30	42	$5\frac{1}{4}$	4	6	4	18	6	15
2929	19	$2\frac{5}{8}$	$18\frac{1}{2}$	$18\frac{1}{4}$	30	42	$17\frac{1}{8}$	$11\frac{3}{4}$	17	$2\frac{1}{4}$	$11\frac{3}{4}$	30	42	$5\frac{1}{4}$	4	6	4	13	6	10
2930	21	$2\frac{5}{8}$	$19\frac{1}{2}$	$19\frac{1}{4}$	30	42	$17\frac{1}{8}$	$11\frac{3}{4}$	19	$2\frac{1}{4}$	$11\frac{3}{4}$	30	42	$5\frac{1}{4}$	4	6	4	15	6	12
2931	21	$2\frac{5}{8}$	$19\frac{1}{2}$	$19\frac{1}{4}$	30	42	$17\frac{5}{8}$	$12\frac{1}{4}$	19	$2\frac{1}{4}$	$11\frac{3}{4}$	30	42	$5\frac{1}{4}$	4	6	4	15	6	12
2932	24	$2\frac{5}{8}$	21	$20\frac{3}{4}$	30	42	$17\frac{1}{2}$	$12\frac{1}{8}$	22	$2\frac{3}{4}$	$11\frac{3}{4}$	30	42	$5\frac{1}{4}$	4	6	4	18	6	15
2933	24	$3\frac{1}{8}$	$21\frac{3}{4}$	$21\frac{3}{4}$	30	42	$17\frac{11}{16}$	$12\frac{1}{4}$	22	$2\frac{3}{4}$	12	30	42	$5\frac{1}{4}$	4	6	4	18	6	15
2934	19	$2\frac{5}{8}$	$18\frac{1}{2}$	$18\frac{1}{4}$	30	42	$17\frac{1}{8}$	$11\frac{3}{4}$	17	$2\frac{1}{4}$	$11\frac{3}{4}$	30	42	$5\frac{1}{4}$	4	6	4	13	6	10
2935	21	$3\frac{1}{8}$	$20\frac{1}{4}$	$20\frac{1}{4}$	30	42	$17\frac{5}{8}$	$12\frac{1}{4}$	19	$2\frac{3}{4}$	12	30	42	$5\frac{1}{4}$	4	6	4	15	6	12
2936	24	$3\frac{1}{8}$	$21\frac{3}{4}$	$21\frac{3}{4}$	30	42	$17\frac{1}{2}$	$12\frac{1}{8}$	22	$2\frac{3}{4}$	12	30	42	$5\frac{1}{4}$	4	6	4	18	6	15
2937	24	$3\frac{1}{8}$	$21\frac{3}{4}$	$21\frac{3}{4}$	30	42	$17\frac{11}{16}$	$12\frac{1}{4}$	22	$2\frac{3}{4}$	12	30	42	$5\frac{1}{4}$	4	6	4	18	6	15

Specifications of Jeffrey Standard Scraper Conveyors using Single Strand of Detachable and Vulcan Chain with Malleable Scrapers
Steel Supports—*For Wood Supports see page 98.*

Length of Conveyor	0 to 50 ft. Centers					51 to 100 ft. Centers					101 to 150 ft. Centers			
No. of Conveyor	2994	2995	2996	2997	2998	2999	3000	3001	3002	3003	3004	3005	3006	3007
Size of Material—Inches														
Average size of Material to be handled	1½	2	2	3	3	1½	2	2	3	3	1½	2	3	3
Maximum size; not to exceed 10% of whole	3	4	4	5	5	3	4	4	5	5	3	4	5	5
Capacity—In tons per hr.														
Horizontal	42	50	50	63	63	42	50	50	63	63	42	50	63	63
15° Incline	23	27	27	33	33	23	27	27	33	33	0	27	33	0
30° Incline	16	20	20	24	24	16	20	20	24	24	0	20	0	0
45° Incline	14	17	17	20	20	14	17	17	20	20	0	0	0	0
Size Scraper—Inches														
Length	10	12	12	15	15	10	12	12	15	15	10	12	15	15
Depth	5	5	5	5	5	5	5	5	5	5	5	5	5	5
Mall. Iron Pattern No.	28026	27898	27898	28656	28656	28026	27898	27898	28656	28656	28026	27898	28656	28656
Spacing—Inches	26	26	24	24.56	24	26	26	24	24.56	24	26	24	24.56	24
Chain														
Number and Style	88J	88J	526V	103J	526V	88J	88J	526V	103J	526V	88J	526V	103J	526V
Pitch—Inches	2.6	2.6	6	3.07	6	2.6	2.6	6.	3.07	6	2.6	6.	3.07	6.
Attachments	F-2	F-2	A-½	F-2	A-½	F-2	F-2	A-½	F-2	A-½	F-2	A-½	F-2	A-½
Working Strength—Lbs	960	960	1640	1600	1640	960	960	1640	1600	1640	960	1640	1600	1640
H. P. At Countershaft*	1.3	1.5	1.8	1.9	2.1	2.6	3.0	3.6	3.8	4.2	3.9	5.3	5.8	6.2
Head Shaft														
Diameter—Inches	1 15/16	1 15/16	1 15/16	1 15/16	2 7/16	2 7/16	2 7/16	2 7/16	2 7/16	2 15/16	2 7/16	2 15/16	2 15/16	2 15/16
Rev. per Min	16⅔	16⅔	16⅔	16⅔	16⅔	16⅔	16⅔	16⅔	16⅔	16⅔	16⅔	16⅔	16⅔	16⅔
Size Sprocket—Inches	23	23	23½	23½	23½	23	23	23½	23½	23½	23	23½	23½	23½
Gear Diam.—Inches	29.83	29.83	29.83	29.83	29.83	29.83	29.83	29.83	29.83	29.83	29.83	29.83	29.83	29.83
Gear Pitch—Inches	1¼	1¼	1¼	1¼	1¼	1¼	1¼	1¼	1¼	1¼	1¼	1¼	1¼	1¼
Gear Face—Inches	3	3	3	3	3	3	3	3	3	3	3	3	3	3
Countershaft														
Diameter—Inches	1 7/16	1 7/16	1 7/16	1 7/16	1 15/16	1 15/16	1 15/16	1 15/16	1 15/16	2 7/16	1 15/16	2 7/16	2 7/16	2 7/16
Rev. per Min	83	83	83	83	83	83	83	83	83	83	83	83	83	83
Pinion Diam.—Inches	6.01	6.01	6.01	6.01	6.01	6.01	6.01	6.01	6.01	6.01	6.01	6.01	6.01	6.01
Pinion Face—Inches	3¼	3¼	3¼	3¼	3¼	3¼	3¼	3¼	3¼	3¼	3¼	3¼	3¼	3¼
Foot Shaft														
Diameter—Inches	1 7/16	1 7/16	1 7/16	1 7/16	1 7/16	1 7/16	1 7/16	1 7/16	1 7/16	1 15/16	1 7/16	1 15/16	1 15/16	1 15/16
Size Sprocket—Inches	23	23	23½	23½	23½	23	23	23½	23½	23½	23	23½	23½	23½
Trough														
Thickness or Gauge	10	10	10	10	3/16	10	10	10	10	3/16	10	10	10	3/16
Approx. Shipping Wgt.—Lbs.														
Terminals, Complete	630	630	680	670	770	730	730	770	760	950	730	930	920	950
Chain and Flights per ft. Ctrs.	10	9	16½	13	17	10	9	16½	13	17	10	16½	13	17
Trough and Bar Trackage per Ft. Ctrs.	10	11	11	12½	16	10	11	11	12½	16	10	11	12½	16

*For Maximum Centers for all inclinations and corresponding capacities.

"Weight of Terminals" given in table includes shafts, bearings, collars, sprockets and gears, with chain and flights half way around the sprockets.

Capacities are figured for 50 lb. material, troughs 80% level full, and uniformly loaded throughout the time period.

For Erection Dimensions of above Conveyors, see page 101.

General Dimensions of Jeffrey Standard Scraper Conveyors

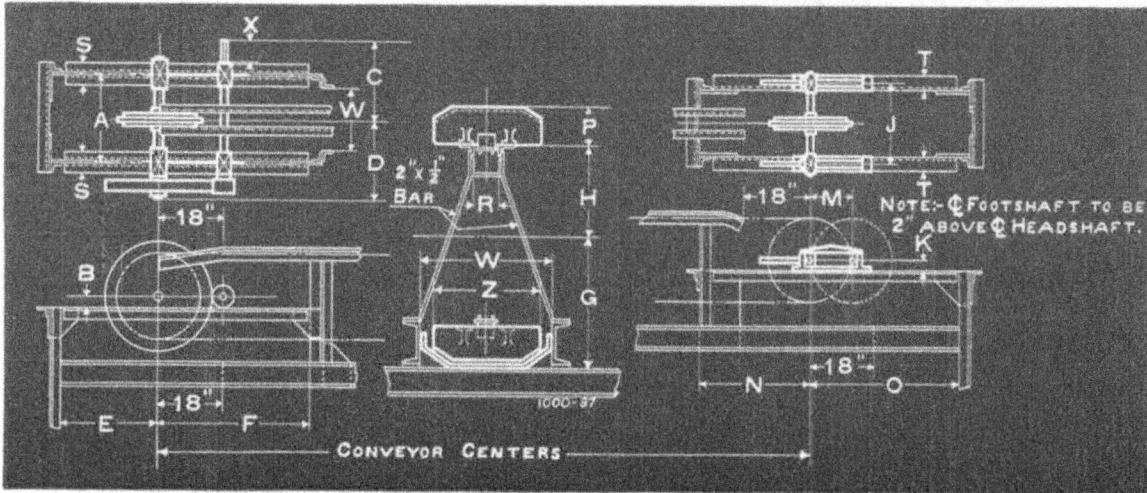

Using Single Strand of Detachable and Vulcan Chain with Malleable Scrapers

Steel Supports—*For Wood Supports see page 99.*   Dimensions in Inches

Conveyor No.	A	B	C	D	E	F	G	H	J	K	M	N	O	P	R	S	T	W	X	Z
2994	19	1¹¹⁄₁₆	17¾	16¾	30	42	17⅛	12	17	2¼	11¾	30	42	5	4	*4	4	13	6	10
2995	21	1¹¹⁄₁₆	18¾	17¾	30	42	17⅛	12	19	2¼	11¾	30	42	5	4	*4	4	15	6	12
2996	21	1¹¹⁄₁₆	18¾	17¾	30	42	17⅝	12½	19	2¼	11¾	30	42	5	4	*4	4	15	6	12
2997	24	1¹¹⁄₁₆	20¼	19¼	30	42	17½	12⅜	22	2¼	11¾	30	42	5	4	*4	4	18	6	15
2998	24	3⅛	21	20¾	30	42	17¹¹⁄₁₆	12½	22	2¼	11¾	30	42	5	4	6¼	4	18	6	15
2999	19	3⅛	18½	18¼	30	42	17⅛	12	17	2¼	11¾	30	42	5	4	6¼	4	13	6	10
3000	21	3⅛	19½	19¼	30	42	17⅛	12	19	2¼	11¾	30	42	5	4	6¼	4	15	6	12
3001	21	3⅛	19½	19¼	30	42	17⅝	12½	19	2¼	11¾	30	42	5	4	6¼	4	15	6	12
3002	24	3⅛	21	20¾	30	42	17½	12⅜	22	2¼	11¾	30	42	5	4	6¼	4	18	6	15
3003	24	3⅝	21¾	21¾	30	42	17¹¹⁄₁₆	12½	22	2¾	12	30	42	5	4	6¼	4	18	6	15
3004	19	3⅛	18½	18¼	30	42	17⅛	12	17	2¼	11¾	30	42	5	4	6¼	4	13	6	10
3005	21	3⅝	20¼	20¼	30	42	17⅝	12½	19	2¾	12	30	42	5	4	6¼	4	15	6	12
3006	24	3⅝	21¾	21¾	30	42	17½	12⅜	22	2¾	12	30	42	5	4	6¼	4	18	6	15
3007	24	3⅝	21¾	21¾	30	42	17¹¹⁄₁₆	12½	22	2¾	12	30	42	5	4	6¼	4	18	6	15

* Use Single Angle for Head Bearing Support.

Specifications of Jeffrey Standard Scraper Conveyors using Single Strand Steel Link and Vulcan Chain, Steel Scrapers with Wearing Blocks

Wood Supports—*For Steel Supports see page 104.*

Lgth. of Conveyor	0 to 50 ft. Centers				51 to 100 ft. Centers				101 to 150 ft. Centers				151 to 200 ft. Centers		
No. of Conveyor	3300	3301	3302	3303	3304	3305	3306	3307	3308	3309	3310	3311	3312	3313	3314
Size of Material—In.															
Avge. size of Material to be handled	1¾	1¾	3	3	1¾	1¾	3	3	1¾	1¾	3	3	1¾	1¾	3
Max. size; not to exceed 10% of whole	3½	3½	5	5	3½	3½	5	5	3½	3½	5	5	3½	3½	5
Capacity—In tons per hr.															
Horizontal	48	48	70	70	48	48	70	70	48	48	70	70	48	48	70
15° Incline	26	26	38	38	26	26	38	38	26	26	0	38	0	26	38
30° Incline	19	19	27	27	19	19	0	27	0	19	0	27	0	19	27
45° Incline	16	16	23	23	16	16	0	23	0	16	0	23	0	16	23
Size Scraper—In.															
Length	15	15	18	18	15	15	18	18	15	15	18	18	15	15	18
Depth	7	7	8	8	7	7	8	8	7	7	8	8	7	7	8
Thickness of Steel	3/16	3/16	1/4	1/4	3/16	3/16	1/4	1/4	3/16	3/16	1/4	1/4	3/16	3/16	1/4
Spacing—Inches	24	32	24	32	24	32	24	32	24	32	24	32	24	32	32
Chain															
Number and Style	526 V	518 S.L.	526 V	518 S.L.	526 V	518 S.L.	526 V	518 S.L.	526 V	518 S.L.	526 V	518 S.L.	526 V	518 S.L.	518 S.L.
Pitch—Inches	6	8	6	8	6	8	6	8	6	8	6	8	6	8	8
Attachments	A½	A1	A½	A1	A½	A1	A½	A1	A½	A1	A½	A1	A½	A1	A1
Work Strength—Lbs.	1640	5200	1640	5200	1640	5200	1640	5200	1640	5200	1640	5200	1640	5200	5200
H. P. at Countershaft*	1.8	1.7	2.4	2.3	3.7	3.5	4.9	4.6	5.5	5.2	7.3	7.0	7.4	7.0	9.3
Head Shaft															
Diameter—In.	2 7/16	2 7/16	2 15/16	2 15/16	2 15/16	2 15/16	2 15/16	3 7/16	2 15/16	3 7/16	2 15/16	3 15/16	2 15/16	3 15/16	4 7/16
Rev. per Min.	16⅔	15	16⅔	15	16⅔	15	16⅔	15	16⅔	15	16⅔	15	16⅔	15	15
Size Sprocket—In.	23½	26	23½	26	23½	26	23½	26	23½	26	23½	26	23½	26	26
Gear Diam.—In.	29.83	29.83	29.83	29.83	29.83	29.83	29.83	29.83	29.83	29.83	40.12	40.12	40.12	40.12	40.12
Gear Pitch—In.	1¼	1¼	1¼	1¼	1¼	1¼	1¼	1¼	1¼	1¼	1½	1½	1½	1½	1½
Gear Face—In.	3	3	3	3	3	3	3	3	3	3	4	4	4	4	4
Countershaft															
Diameter—In.	1 15/16	1 15/16	2 7/16	2 7/16	2 7/16	2 7/16	2 7/16	2 11/16	2 7/16	2 11/16	2 7/16	2 15/16	2 7/16	2 15/16	3 7/16
Rev. per Min.	83	75	83	75	83	75	83	75	83	75	93	84	93	84	84
Pinion Diam.—In.	6.01	6.01	6.01	6.01	6.01	6.01	6.01	6.01	6.01	6.01	7.22	7.22	7.22	7.22	7.22
Pinion Face—In.	3¼	3¼	3¼	3¼	3¼	3¼	3¼	3¼	3¼	3¼	4½	4½	4½	4½	4½
Foot Shaft															
Diameter—In.	1 7/16	1 7/16	1 15/16	1 15/16	1 15/16	1 15/16	1 15/16	2 7/16	1 15/16	2 7/16	1 15/16	2 7/16	1 15/16	2 7/16	2 15/16
Size Sprocket—In.	23½	26	23½	26	23½	26	23½	26	23½	26	23½	26	23½	26	26
Trough															
Thickness or Gauge	10	3/16	10	3/16	10	3/16	10	3/16	10	3/16	10	3/16	10	3/16	3/16
Approx. Shipping Wgt.—Lbs.															
Terminals, Complete	830	940	1040	1145	1020	1120	1040	1310	1020	1280	1280	1680	1230	1580	2120
Chain and Flights per Ft. Ctrs.	24	20½	28	24½	24	20½	28	24½	24	20½	28	24½	24	20½	24½
Trough and Bar Trackage per Ft. Ctrs.	18½	22½	21	25½	18½	22½	21	25½	18½	22½	21	25½	18½	22½	25½

*For Maximum Centers for all inclinations and corresponding capacities.

"Weight of Terminals" given in table includes shafts, bearings, collars, sprockets and gears, with chain and flights half way around the sprockets.

Capacities are figured for 50 lb. material, troughs 80% level full, and uniformly loaded throughout the time period.

For Erection Dimensions of above Conveyors, see page 103.

General Dimensions of Jeffrey Standard Scraper Conveyors

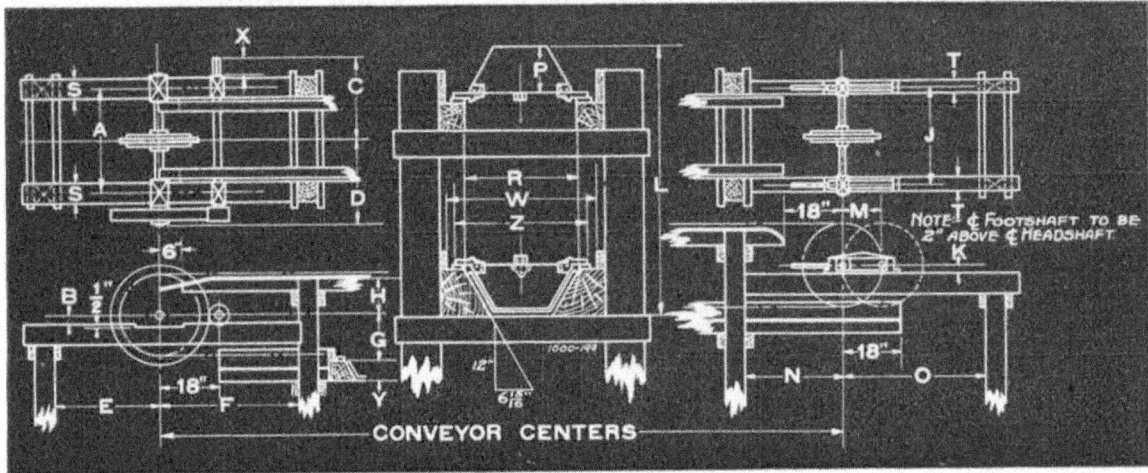

Using Single Strand Steel Link and Vulcan Chain, Steel Scrapers with Wearing Blocks.

Wood Supports—*For Steel Supports see page 105.* **Dimensions in Inches.**

Conveyor No.	A	B	C	D	E	F	G	H	J	K	L	M	N	O	P	R	S	T	W	X	Y	Z
3300	28½	2⅝	23¼	23	32	42	12¾	11¼	26½	2¼	39⅜	11¾	30	44	7	16 5/16	6	4	20½	6	7⅛	19
3301	28½	2⅝	23¼	23	34	42	13½	13¼	26½	2¼	43½	11¾	30	46	7	16⅜	6	4	20½	6	7 3/16	19
3302	31½	3⅛	25½	25½	32	42	12¾	11¼	29½	2¾	41⅜	12	30	44	8	19 5/16	6	4	23½	6	8⅛	22
3303	31½	3⅛	25½	25½	34	42	13½	13¼	29½	2¾	45½	12	30	46	8	19 7/16	6	4	23½	6	8 3/16	22
3304	28½	3⅛	24	24	32	42	12¾	11¼	26½	2¾	39⅜	12	30	44	7	16 5/16	6	4	20½	6	7⅛	19
3305	28½	3⅛	24	24	34	42	13½	13¼	26½	2¾	43½	12	30	46	7	16⅜	6	4	20½	6	7 3/16	19
3306	31½	3⅛	25½	25½	32	42	12¾	11¼	29½	2¾	41⅜	12	30	44	8	19 5/16	6	4	23½	6	8⅛	22
3307	33½	3½	27⅞	27½	34	42	13½	13¼	31½	3⅛	45½	15	32	49	8	19 7/16	8	6	23½	7	8 3/16	22
3308	28½	3⅛	24	24	32	42	12¾	11¼	26½	2¾	39⅜	12	30	44	7	16 5/16	6	4	20½	6	7⅛	19
3309	30½	3½	26⅜	26	34	42	13½	13¼	28½	3⅛	43½	15	32	49	7	16⅜	8	6	20½	7	7 3/16	19
3310	31½	3⅜	25½	25½	32	48	12¾	11¼	29½	2¾	41⅜	12	30	44	8	19 5/16	6	4	23½	6	8⅛	22
3311	33½	4⅛	29¼	29	34	48	13½	13¼	31½	3⅛	45½	15	32	49	8	19 7/16	8	6	23½	8	8 3/16	22
3312	28½	3⅜	24	24	32	48	12¾	11¼	26½	2¾	39⅜	12	30	44	7	16 5/16	6	4	20½	6	7⅛	19
3313	30½	4⅜	27¾	27½	34	48	13½	13¼	28½	3⅛	43½	15	32	49	7	16⅜	8	6	20½	8	7 3/16	19
3314	35½	4⅞	32	31½	34	48	13½	13¼	31½	4	45½	22¼	44	56	8	19 7/16	10	6	23½	9	8 3/16	22

Specifications of Jeffrey Standard Scraper Conveyors using Single Strand Steel Link and Vulcan Chain, Steel Scrapers with Wearing Blocks

Steel Supports—*For Wood Supports see page 102.*

Length of Conveyor	0 to 50 ft. Centers				51 to 100 ft. Centers				101 to 150 ft. Centers				151 to 200ft. Centers		
No. of Conveyor	3315	3316	3317	3318	3319	3320	3321	3322	3323	3324	3325	3326	3327	3328	3329
Size of Material—In.															
Avge. size of Material to be handled	$1\frac{3}{4}$	$1\frac{3}{4}$	3	3	$1\frac{3}{4}$	$1\frac{3}{4}$	3	3	$1\frac{3}{4}$	$1\frac{3}{4}$	3	3	$1\frac{3}{4}$	$1\frac{3}{4}$	3
Max. size; not to exceed 10% of whole	$3\frac{1}{2}$	$3\frac{1}{2}$	5	5	$3\frac{1}{2}$	$3\frac{1}{2}$	5	5	$3\frac{1}{2}$	$3\frac{1}{2}$	5	5	$3\frac{1}{2}$	$3\frac{1}{2}$	5
Capacity—In tons per hr.															
Horizontal	48	48	70	70	48	48	70	70	48	48	70	70	48	48	70
15° Incline	26	26	38	38	26	26	38	38	26	26	0	38	0	26	38
30° Incline	19	19	27	27	19	19	0	27	0	19	0	27	0	19	27
45° Incline	16	16	23	23	16	16	0	23	0	16	0	23	0	16	23
Size Scraper—In.															
Length	15	15	18	18	15	15	18	18	15	15	18	18	15	15	18
Depth	7	7	8	8	7	7	8	8	7	7	8	8	7	7	8
Thickness of Steel	$\frac{3}{16}$	$\frac{3}{16}$	$\frac{1}{4}$	$\frac{1}{4}$	$\frac{3}{16}$	$\frac{3}{16}$	$\frac{1}{4}$	$\frac{1}{4}$	$\frac{3}{16}$	$\frac{3}{16}$	$\frac{1}{4}$	$\frac{1}{4}$	$\frac{3}{16}$	$\frac{3}{16}$	$\frac{1}{4}$
Spacing—Inches	24	32	24	32	24	32	24	32	24	32	24	32	24	32	32
Chain															
Number and Style	526 V	518 S.L	526 V	518 S.L	526 V	518 S.L	526 V	518 S.L	526 V	518 S.L	526 V	518 S.L	526 V	518 S.L	518 S.L
Pitch—Inches	6	8	6	8	6	8	6	8	6	8	6	8	6	8	8
Attachments	$A\frac{1}{2}$	A1	$A\frac{1}{2}$	A1	$A\frac{1}{2}$	A1	$A\frac{1}{2}$	A1	$A\frac{1}{2}$	A1	$A\frac{1}{2}$	A1	$A\frac{1}{2}$	A1	A1
Work. Strength–Lbs.	1640	5200	1640	5200	1640	5200	1640	5200	1640	5200	1640	5200	1640	5200	5200
H. P. at Countershaft*	1.8	1.7	2.4	2.3	3.7	3.5	4.9	4.6	5.5	5.2	7.3	7.0	7.4	7.0	9.3
Head Shaft															
Diameter—Inches	$2\frac{7}{16}$	$2\frac{7}{16}$	$2\frac{15}{16}$	$2\frac{15}{16}$	$2\frac{15}{16}$	$2\frac{15}{16}$	$2\frac{15}{16}$	$3\frac{7}{16}$	$2\frac{15}{16}$	$3\frac{7}{16}$	$2\frac{15}{16}$	$3\frac{15}{16}$	$2\frac{15}{16}$	$3\frac{15}{16}$	$4\frac{7}{16}$
Rev. per Min.	$16\frac{2}{3}$	15	$16\frac{2}{3}$	15	$16\frac{2}{3}$	15	$16\frac{2}{3}$	15	$16\frac{2}{3}$	15	$16\frac{2}{3}$	15	$16\frac{2}{3}$	15	15
Size Sprocket—In.	$23\frac{1}{2}$	26	$23\frac{1}{2}$	26	$23\frac{1}{2}$	26	$23\frac{1}{2}$	26	$23\frac{1}{2}$	26	$23\frac{1}{2}$	26	$23\frac{1}{2}$	26	26
Gear Diam.—In.	29.83	29.83	29.83	29.83	29.83	29.83	29.83	29.83	29.83	29.83	40.12	40.12	40.12	40.12	40.12
Gear Pitch—In.	$1\frac{1}{4}$	$1\frac{1}{4}$	$1\frac{1}{4}$	$1\frac{1}{4}$	$1\frac{1}{4}$	$1\frac{1}{4}$	$1\frac{1}{4}$	$1\frac{1}{4}$	$1\frac{1}{4}$	$1\frac{1}{4}$	$1\frac{1}{2}$	$1\frac{1}{2}$	$1\frac{1}{2}$	$1\frac{1}{2}$	$1\frac{1}{2}$
Gear Face—In.	3	3	3	3	3	3	3	3	3	3	4	4	4	4	4
Countershaft															
Diameter—In.	$1\frac{15}{16}$	$1\frac{15}{16}$	$2\frac{7}{16}$	$2\frac{7}{16}$	$2\frac{7}{16}$	$2\frac{7}{16}$	$2\frac{7}{16}$	$2\frac{11}{16}$	$2\frac{7}{16}$	$2\frac{11}{16}$	$2\frac{7}{16}$	$2\frac{15}{16}$	$2\frac{7}{16}$	$2\frac{15}{16}$	$3\frac{7}{16}$
Rev. per Min.	83	75	83	75	83	75	83	75	83	75	83	84	84	84	84
Pinion Diam.—In.	6.01	6.01	6.01	6.01	6.01	6.01	6.01	6.01	6.01	6.01	7.22	7.22	7.22	7.22	7.22
Pinion Face—In.	$3\frac{1}{4}$	$3\frac{1}{4}$	$3\frac{1}{4}$	$3\frac{1}{4}$	$3\frac{1}{4}$	$3\frac{1}{4}$	$3\frac{1}{4}$	$3\frac{1}{4}$	$3\frac{1}{4}$	$3\frac{1}{4}$	$4\frac{1}{2}$	$4\frac{1}{2}$	$4\frac{1}{2}$	$4\frac{1}{2}$	$4\frac{1}{2}$
Foot Shaft															
Diameter—In.	$1\frac{7}{16}$	$1\frac{7}{16}$	$1\frac{15}{16}$	$1\frac{15}{16}$	$1\frac{15}{16}$	$1\frac{15}{16}$	$1\frac{15}{16}$	$2\frac{7}{16}$	$1\frac{15}{16}$	$2\frac{7}{16}$	$1\frac{15}{16}$	$2\frac{7}{16}$	$1\frac{15}{16}$	$2\frac{7}{16}$	$2\frac{15}{16}$
Size Sprocket—In.	$23\frac{1}{2}$	26	$23\frac{1}{2}$	26	$23\frac{1}{2}$	26	$23\frac{1}{2}$	26	$23\frac{1}{2}$	26	$23\frac{1}{2}$	26	$23\frac{1}{2}$	26	26
Trough															
Thickness or Gauge	10	$\frac{3}{16}$	10	$\frac{3}{16}$	10	$\frac{3}{16}$	10	$\frac{3}{16}$	10	$\frac{3}{16}$	10	$\frac{3}{16}$	10	$\frac{3}{16}$	$\frac{3}{16}$
Approx. Shipping Wgt.—Lbs.															
Terminals Complete	830	940	1040	1145	1020	1120	1040	1310	1020	1280	1280	1680	1230	1580	2120
Chain and Flights per Ft. Ctrs.	24	$20\frac{1}{2}$	28	$24\frac{1}{2}$	24	$20\frac{1}{2}$	28	$24\frac{1}{2}$	24	$20\frac{1}{2}$	28	$24\frac{1}{2}$	24	$20\frac{1}{2}$	$24\frac{1}{2}$
Trough and Bar Trackage per Ft. Ctrs.	$18\frac{1}{2}$	$22\frac{1}{2}$	21	$25\frac{1}{2}$	$18\frac{1}{2}$	$22\frac{1}{2}$	21	$25\frac{1}{2}$	$18\frac{1}{2}$	$22\frac{1}{2}$	21	$25\frac{1}{2}$	$18\frac{1}{2}$	$22\frac{1}{2}$	$25\frac{1}{2}$

* For Maximum Centers for all inclinations and corresponding capacities.

"Weight of Terminals" given in table includes shafts, bearings, collars, sprockets and gears, with chain and flights half way around the sprockets.

Capacities are figured for 50 lb. material, troughs 80% level full, and uniformly loaded throughout the time period.

For Erection Dimensions of above Conveyors, see page 105.

General Dimensions of Jeffrey Standard Scraper Conveyors

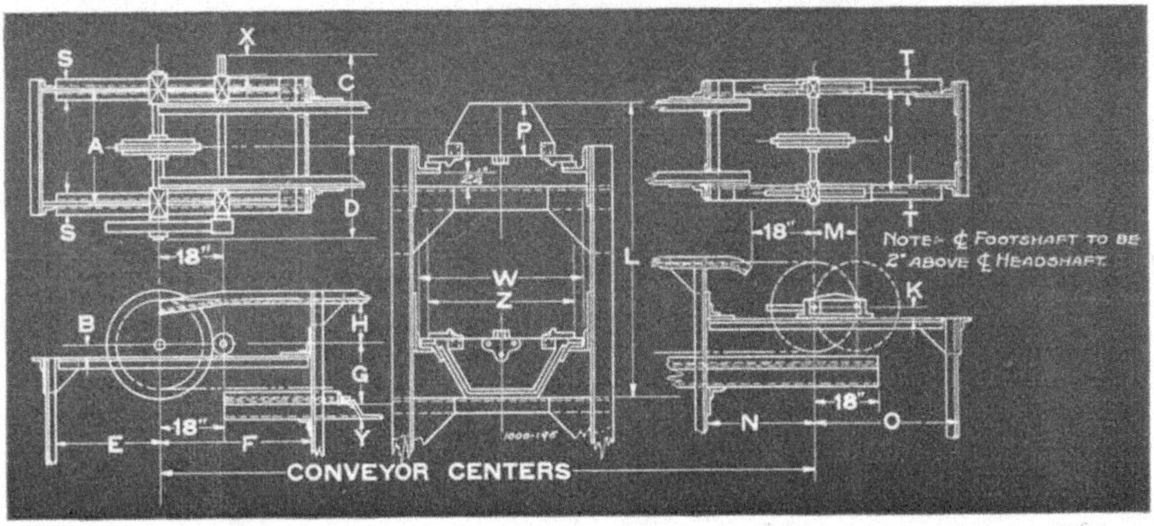

Using Single Strand Steel Link and Vulcan Chain, Steel Scrapers with Wearing Blocks

Steel Supports—*For Wood Supports* *see page 103.* **Dimensions in Inches**

Conveyor No.	A	B	C	D	E	F	G	H	J	K	L	M	N	O	P	S	T	W	X	Y	Z
3315	28½	3⅛	23¼	23	32	42	13⅛	11	26½	2¼	39⅜	11¾	30	44	7	6¼	4	20¾	6	6¾	19
3316	28½	3⅛	23¼	23	34	42	15 3/16	13	26½	2¼	43½	11¾	30	46	7	6¼	4	20¾	6	6¾	19
3317	31½	3⅝	25½	25½	32	42	13⅛	11	29½	2¾	41⅜	12	30	44	8	6¼	4	23¾	6	7¾	22
3318	31½	3⅝	25½	25½	34	42	15 3/16	13	29½	2¾	45½	12	30	46	8	6¼	4	23¾	6	7¾	22
3319	28½	3⅝	24	24	32	42	13⅛	11	26½	2¾	39⅜	12	30	44	7	6¼	4	20¾	6	6¾	19
3320	28½	3⅝	24	24	34	42	15 3/16	13	26½	2¾	43½	12	30	46	7	6¼	4	20¾	6	6¾	19
3321	31½	3⅝	25½	25½	32	42	13⅛	11	29½	2¾	41⅜	12	30	44	8	6¼	4	23¾	6	7¾	22
3322	33½	4	27⅞	27½	34	42	15 3/16	13	29½	3⅛	45½	15	32	49	8	7¼	4	23¾	7	7¾	22
3323	28½	3⅝	24	24	32	42	13⅛	11	26½	2¾	39⅜	12	30	44	7	6¼	4	20¾	6	6¾	19
3324	30½	4	26⅜	26	34	42	15 3/16	13	26½	3⅛	43½	15	32	49	7	7¼	4	20¾	7	6¾	19
3325	31½	3⅝	25½	25½	32	48	13⅛	11	29½	2¾	41⅜	12	30	44	8	6¼	4	23¾	6	7¾	22
3326	33½	4⅝	29¼	29	34	48	15 3/16	13	29½	3⅛	45½	15	32	49	8	8¼	4	23¾	8	7¾	22
3327	28½	3⅝	24	24	32	48	13⅛	11	26½	2¾	39⅜	12	30	44	7	6¼	4	20¾	6	6¾	19
3328	30½	4⅝	27¾	27½	34	48	15 3/16	13	26½	3⅛	43½	15	32	49	7	8¼	4	20¾	8	6¾	19
3329	35½	5⅜	32	31½	34	48	15 3/16	13	31¾	4	45½	22¼	44	56	8	10¼	*6¼	23¾	9	7¾	22

*Use Double Angle for 2 15/16″ Takeups.

Specifications of Jeffrey Standard Scraper Conveyors Using Single Strand Steel Link and Vulcan Chain, Steel Scrapers with Roller Attachments

Wood Supports—*For Steel Supports see page 108.*

Lgth. of Conveyor	0 to 50 ft. Centers				51 to 100 ft. Centers				101 to 150 ft. Centers				151 to 200 ft. Centers		
No. of Conveyor	2938	2939	2940	2941	2942	2943	2944	2945	2946	2947	2948	2949	2950	2951	2952
Size of Material—In.															
Avge. size of Material to be handled	1¾	1¾	3	3	1¾	1¾	3	3	1¾	1¾	3	3	1¾	1¾	3
Max. size; not to exceed 10% of whole	3½	3½	5	5	3½	3½	5	5	3½	3½	5	5	3½	3½	5
Capacity—In tons per hr.															
Horizontal	48	48	70	70	48	48	70	70	48	48	70	70	48	48	70
15° Incline	26	26	38	38	26	26	38	38	26	26	0	38	0	26	38
30° Incline	19	19	27	27	19	19	0	27	0	19	0	27	0	19	27
45° Incline	16	16	23	23	16	16	0	23	0	16	0	23	0	16	23
Size Scraper—In.															
Length	15	15	18	18	15	15	18	18	15	15	18	18	15	15	18
Depth	7	7	8	8	8	7	8	8	7	7	8	8	7	7	8
Thickness of Steel	3/16	3/16	¼	¼	3/16	3/16	¼	¼	3/16	3/16	¼	¼	3/16	3/16	¼
Spacing, Inches	24	32	24	32	24	32	24	32	24	32	24	32	24	32	32
Chain															
Number and Style	526 V	518 S.L.	526 V	518 S.L.	526 V	518 S.L.	526 V	518 S.L.	526 V	518 S.L.	526 V	518 S.L.	526 V	518 S.L.	518 S.L.
Pitch—Inches	6	8	6	8	6	8	6	8	6	8	6	8	6	8	8
Attachments	A½	A1	A½	A1	A½	A1	A½	A1	A½	A1	A½	A1	A½	A1	A1
Work. Strength-Lbs.	1640	5200	1640	5200	1640	5200	1640	5200	1640	5200	1640	5200	1640	5200	5200
H. P. At Countershaft*	1.8	1.7	2.4	2.3	3.6	3.4	4.6	4.6	5.1	5.1	6.7	6.9	6.7	6.8	9.2
Head Shaft															
Diameter—In.	2 7/16	2 7/16	2 15/16	2 15/16	2 15/16	2 15/16	2 15/16	3 7/16	2 15/16	3 7/16	2 15/16	3 15/16	2 15/16	3 15/16	4 7/16
Rev. per Min.	16⅔	15	16⅔	15	16⅔	15	16⅔	15	16⅔	15	16⅔	15	16⅔	15	15
Size Sprocket—In.	23½	26	23½	26	23½	26	23½	26	23½	26	23½	26	23½	26	26
Gear Diam.—In.	29.83	29.83	29.83	29.83	29.83	29.83	29.83	29.83	29.83	29.83	40.12	40.12	40.12	40.12	40.12
Gear Pitch—In.	1¼	1¼	1¼	1¼	1¼	1¼	1¼	1¼	1¼	1¼	1½	1½	1½	1½	1½
Gear Face—In.	3	3	3	3	3	3	3	3	3	3	4	4	4	4	4
Countershaft															
Diameter—In.	1 15/16	1 15/16	2 7/16	2 7/16	2 7/16	2 7/16	2 7/16	2 11/16	2 7/16	2 11/16	2 7/16	2 15/16	2 7/16	2 15/16	3 7/16
Rev. per Min.	83	75	83	75	83	75	83	75	83	75	93	84	93	84	84
Pinion Diam.—In.	6.01	6.01	6.01	6.01	6.01	6.01	6.01	6.01	6.01	6.01	7.22	7.22	7.22	7.22	7.22
Pinion Face—In.	3¼	3¼	3¼	3¼	3¼	3¼	3¼	3¼	3¼	3¼	4½	4½	4½	4½	4½
Foot Shaft															
Diameter—In.	1 7/16	1 7/16	1 15/16	1 15/16	1 15/16	1 15/16	1 15/16	2 7/16	1 15/16	2 7/16	1 15/16	2 7/16	1 15/16	2 7/16	2 15/16
Size Sprocket—In.	23½	26	23½	26	23½	26	23½	26	23½	26	23½	26	23½	26	26
Trough															
Thickness or Gauge	10	3/16	10	3/16	10	3/16	10	3/16	10	3/16	10	3/16	10	3/16	4/16
Approx. Shipping Wgt.—Lbs.															
Terminals, Complete	830	940	1040	1145	1020	1120	1040	1310	1020	1280	1280	1680	1230	1580	2120
Chain and Flights per Ft. Ctrs.	30	26	34	30	30	26	34	30	30	26	34	30	30	26	30
Trough and Bar Trackage per Ft. Ctrs.	18½	22½	21	25½	18½	22½	21	25½	18½	22½	21	25½	18½	22½	25½

*For Maximum Centers for all inclinations and corresponding capacities.

"Weight of Terminals" given in table includes shafts, bearings, collars, sprockets and gears, with chain and flights half way around the sprockets.

Capacities are figured for 50 lb. material, troughs 80% level full, and uniformly loaded throughout the time period.

For Erection Dimensions of above Conveyors, see page 107.

General Dimensions of Jeffrey Standard Scraper Conveyors

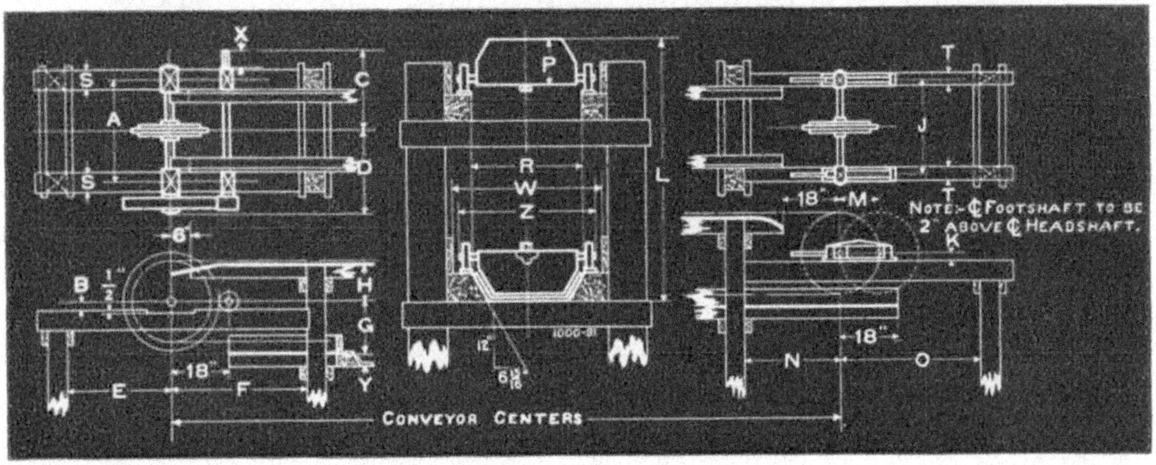

Using Single Strand Steel Link and Vulcan Chain, Steel Scrapers with Roller Attachments

Wood Supports—*For Steel Supports see page 109.* **Dimensions in Inches.**

Conveyor No.	A	B	C	D	E	F	G	H	J	K	L	M	N	O	P	R	S	T	W	X	Y	Z
2938	31	2⅝	24½	24¼	32	42	15¾	11¼	29	2¼	39⅜	11¾	30	44	7	17	6	4	23	6	4⅛	21½
2939	31	2⅝	24½	24¼	34	42	17¾	13¼	29	2¼	43½	11¾	30	46	7	17	6	4	23	6	4⅛	21½
2940	34	3⅛	26¾	26¾	32	42	15¾	11¼	32	2¾	41⅜	12	30	44	8	20	6	4	26	6	5$\frac{1}{16}$	24½
2941	34	3⅛	26¾	26¾	34	42	17¾	13¼	32	2¾	45½	12	30	46	8	20	6	4	26	6	5$\frac{1}{16}$	24½
2942	31	3⅛	25¼	25¼	32	42	15¾	11¼	29	2¾	39⅜	12	30	44	7	17	6	4	23	6	4⅛	21½
2943	31	3⅛	25¼	25¼	34	42	17¾	13¼	29	2¾	43½	12	30	46	7	17	6	4	23	6	4⅛	21½
2944	34	3⅛	26¾	26¾	32	42	15¾	11¼	32	2¾	41⅜	12	30	44	8	20	6	4	26	6	5$\frac{1}{16}$	24½
2945	36	3½	29⅛	28¾	34	42	17¾	13¼	34	3⅛	45½	15	32	49	8	20	8	6	26	7	5$\frac{1}{16}$	24½
2946	31	3⅛	25¼	25¼	32	42	15¾	11¼	29	2¾	39⅜	12	30	44	7	17	6	4	23	6	4⅛	21½
2947	33	3½	27⅝	27¼	34	42	17¾	13¼	31	3⅛	43½	15	32	49	7	17	8	6	23	7	4⅛	21½
2948	34	3⅛	26¾	26¾	32	48	15¾	11¼	32	2¾	41⅜	12	30	44	8	20	6	4	26	6	5$\frac{1}{16}$	24½
2949	36	4⅛	30½	30¼	34	48	17¾	13¼	34	3⅛	45½	15	32	49	8	20	8	6	26	8	5$\frac{1}{16}$	24½
2950	31	3⅛	25¼	25¼	32	48	15¾	11¼	29	2¾	39⅜	12	30	44	7	17	6	4	23	6	4⅛	21½
2951	33	4⅛	29	28¾	34	48	17¾	13¼	31	3⅛	43½	15	32	49	7	17	8	6	23	8	4⅛	21½
2952	38	4⅞	33¼	32¾	34	48	17¾	13¼	34	4	45½	22¼	44	56	8	20	10	6	26	9	5$\frac{1}{16}$	24½

Scraper Conveyors ———— JEFFREY

Specifications of Jeffrey Standard Scraper Conveyors using Single Strand Steel Link and Vulcan Chain, Steel Scrapers with Roller Attachments

Steel Supports—*For Wood Supports see page 106.*

Length of Conveyor	0 to 50 ft. Centers				51 to 100 ft. Centers				101 to 150 ft. Centers				151to200ft.Centers		
No. of Conveyor	3008	3009	3010	3011	3012	3013	3137	3138	3139	3140	3141	3142	3143	3144	3145
Size of Material—In.															
Avge. size of Material to be handled	1¾	1¾	3	3	1¾	1¾	3	3	1¾	1¾	3	3	1¾	1¾	3
Max. size; not to exceed 10% of whole	3½	3½	5	5	3½	3½	5	5	3½	3½	5	5	3½	3½	5
Capacity—In tons per hr.															
Horizontal	48	48	70	70	48	48	70	70	48	48	70	70	48	48	70
15° Incline	26	26	38	38	26	26	38	38	26	26	0	38	0	26	38
30° Incline	19	19	27	27	19	19	0	27	0	19	0	27	0	19	27
45° Incline	16	16	23	23	16	16	0	23	0	16	0	23	0	16	23
Size Scraper—In.															
Length	15	15	18	18	15	15	18	18	15	15	18	18	15	15	18
Depth	7	7	8	8	7	7	8	8	7	7	8	8	7	7	8
Thickness of Steel	3/16	3/16	1/4	1/4	3/16	3/16	1/4	1/4	3/16	3/16	1/4	1/4	3/16	3/16	1/4
Spacing—Inches	24	32	24	32	24	32	24	32	24	32	24	32	24	32	32
Chain															
Number and Style	526 V	518 S.L	526 V	518 S.L	526 V	518 S.L	526 V	518 S.L	526 V	518 S.L	526 V	518 S.L	526 V	518 S.L	518 S.L
Pitch—Inches	6	8	6	8	6	8	6	8	6	8	6	8	6	8	8
Attachments	A½	A1	A½	A1	A½	A1	A½	A1	A½	A1	A½	A1	A½	A1	A1
Work. Strength–Lbs.	1640	5200	1640	5200	1640	5200	1640	5200	1640	5200	1640	5200	1640	5200	5200
H. P. at Countershaft*	1.8	1.7	2.4	2.3	3.6	3.4	4.6	4.6	5.1	5.1	6.7	6.9	6.7	6.8	9.2
Head Shaft															
Diameter—Inches	2 7/16	2 7/16	2 15/16	2 15/16	2 15/16	2 15/16	2 15/16	3 7/16	2 15/16	3 7/16	2 15/16	3 15/16	2 15/16	3 15/16	4 7/16
Rev. per Min.	16⅔	15	16⅔	15	16⅔	15	16⅔	15	16⅔	15	16⅔	15	16⅔	15	15
Size Sprocket—In.	23½	26	23½	26	23½	26	23½	26	23½	26	23½	26	23½	26	26
Gear Diam.—In.	29.83	29.83	29.83	29.83	29.83	29.83	29.83	29.83	29.83	29.83	40.12	40.12	40.12	40.12	40.12
Gear Pitch—In.	1¼	1¼	1¼	1¼	1¼	1¼	1¼	1¼	1¼	1¼	1½	1½	1½	1½	1½
Gear Face—In.	3	3	3	3	3	3	3	3	3	3	4	4	4	4	4
Countershaft															
Diameter—In.	1 15/16	1 15/16	2 7/16	2 7/16	2 7/16	2 7/16	2 7/16	2 11/16	2 7/16	2 11/16	2 7/16	2 15/16	2 7/16	2 15/16	3 7/16
Rev. per Min.	83	75	83	75	83	75	83	75	83	75	93	84	93	84	84
Pinion Diam.—In.	6.01	6.01	6.01	6.01	6.01	6.01	6.01	6.01	6.01	6.01	7.22	7.22	7.22	7.22	7.22
Pinion Face—In.	3¼	3¼	3¼	3¼	3¼	3¼	3¼	3¼	3¼	3¼	4½	4½	4½	4½	4½
Foot Shaft															
Diameter—In.	1 7/16	1 7/16	1 15/16	1 15/16	1 15/16	1 15/16	1 15/16	2 7/16	1 15/16	2 7/16	1 15/16	2 7/16	1 15/16	2 7/16	2 15/16
Size Sprocket—In.	23½	26	23½	26	23½	26	23½	26	23½	26	23½	26	23½	26	26
Trough															
Thickness or Gauge	10	3/16	10	3/16	10	3/16	10	3/16	10	3/16	10	3/16	10	3/16	3/16
Approx. Shipping Wgt.—Lbs.															
Terminals Complete	830	940	1040	1145	1020	1120	1040	1310	1020	1280	1280	1680	1230	1580	2120
Chain and Flights per Ft. Ctrs.	30	26	34	30	30	26	34	30	30	26	34	30	30	26	30
Trough and Bar Trackage per Ft. Ctrs.	21	26	23½	28	21	26	23½	28	21	26	23½	28	21	26	28

*For Maximum Centers for all inclinations and corresponding capacities.

"Weight of Terminals" given in table includes shafts, bearings, collars, sprockets and gears, with chain and flights half way around the sprockets.

Capacities are figured for 50 lb. material, troughs 80% level full, and uniformly loaded throughout the time period.
For Erection Dimensions of above Conveyors, see page 109.

General Dimensions of Jeffrey Standard Scraper Conveyors

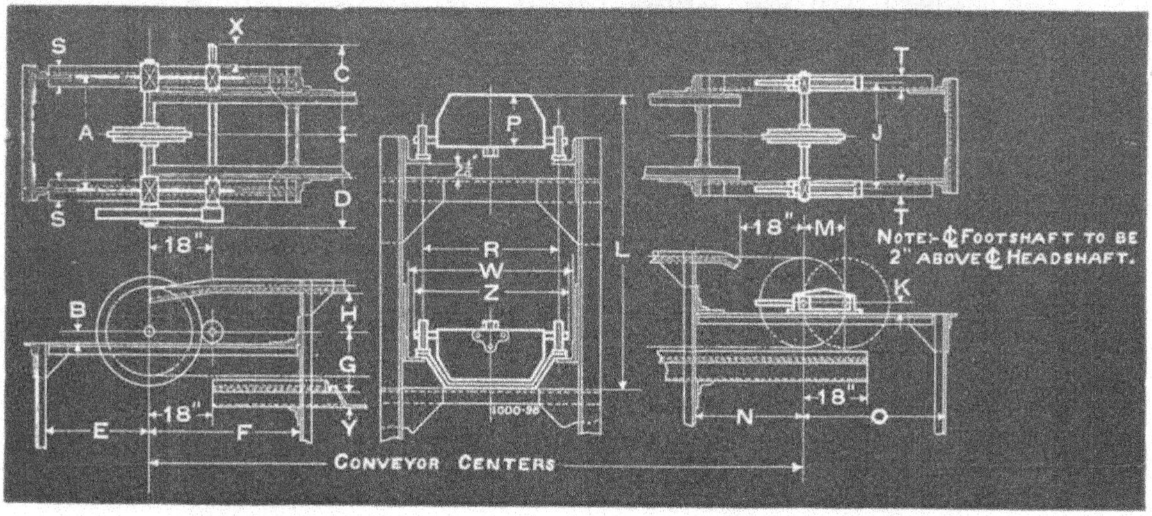

Using Single Strand Steel Link and Vulcan Chain, Steel Scrapers with Roller Attachments

Steel Supports—*For Wood Supports see page 107.* **Dimensions in Inches**

Conveyor No.	A	B	C	D	E	F	G	H	J	K	L	M	N	O	P	R	S	T	W	X	Y	Z
3008	31	3⅛	24½	24¼	32	42	16⅛	11	29	2¼	39⅜	11¾	30	44	7	19	6¼	4	23	6	3¾	21½
3009	31	3⅛	24½	24¼	34	42	18 3/16	13	29	2¼	43½	11¾	30	46	7	19	6¼	4	23	6	3¾	21½
3010	34	3⅝	26¾	26¾	32	42	16⅛	11	32	2¾	41⅜	12	30	44	8	22	6¼	4	26	6	4¾	24½
3011	34	3⅝	26¾	26¾	34	42	18 3/16	13	32	2¾	45½	12	30	46	8	22	6¼	4	26	6	4¾	24½
3012	31	3⅝	25¼	25¼	32	42	16⅛	11	29	2¾	39⅜	12	30	44	7	19	6¼	4	23	6	3¾	21½
3013	31	3⅝	25¼	25¼	34	42	18 3/16	13	29	2¾	43½	12	30	46	7	19	6¼	4	23	6	3¾	21½
3137	34	3⅝	26¾	26¾	32	42	16⅛	11	32	2¾	41⅜	12	30	44	8	22	6¼	4	26	6	4¾	24½
3138	36	4	29⅛	28¾	34	42	18 3/16	13	34	3⅛	45½	15	32	49	8	22	7¼	4	26	7	4¾	24½
3139	31	3⅝	25¼	25¼	32	42	16⅛	11	29	2¾	39⅜	12	30	44	7	19	6¼	4	23	6	3¾	21½
3140	33	4	27⅝	27¼	34	42	18 3/16	13	31	3⅛	43½	15	32	49	7	19	7¼	4	23	7	3¾	21½
3141	34	3⅝	26¾	26¾	32	48	16⅛	11	32	2¾	41⅜	12	30	44	8	22	6¼	4	26	6	4¾	24½
3142	36	4⅝	30½	30¼	34	48	18 3/16	13	34	3⅛	45½	15	32	49	8	22	8¼	4	26	8	4¾	24½
3143	31	3⅝	25¼	25¼	32	48	16⅛	11	29	2¾	39⅜	12	30	44	7	19	6¼	4	23	6	3¾	21½
3144	33	4⅝	29	28¾	34	48	18 3/16	13	31	3⅛	43½	15	32	49	7	19	8¼	4	23	8	3¾	21½
3145	38	5⅜	33¼	32¾	34	48	18 3/16	13	34	4	45½	22¼	44	56	8	22	10¼	*6¼	26	9	4¾	24½

*Use Double Angles for 2 15/16" Takeups.

Specifications of Jeffrey Standard Scraper Conveyors using Double Strand Malleable Roller Chain with Steel Scrapers

Wood Supports—*For Steel Supports see page 112.*

Length of Conveyor	0 to 50 ft. Centers			51 to 100 ft. Centers			101 to 150 ft. Centers			151 to 200 ft. Centers		
No. of Conveyor	2953	2954	2955	2956	2957	2958	2959	2960	2961	2962	2963	2964
Size of Material—In.												
Avge. size of Material to be handled	6	6	8	6	6	8	6	6	8	6	6	8
Max. size; not to exceed 10% of whole..	9	9	12	9	9	12	9	9	12	9	9	12
Capacity—In tons per hr.												
Horizontal	60	60	112	60	60	112	60	60	112	60	60	112
15° Incline	32	32	60	32	32	60	32	32	60	32	32	60
30° Incline	23	23	44	23	23	44	23	23	44	23	23	44
45° Incline	20	20	37	20	20	37	20	20	37	20	20	37
Size Scraper—In.												
Length	18	18	24	18	18	24	18	18	24	18	18	24
Depth	6	6	8	6	6	8	6	6	8	6	6	8
Thickness of Steel	¼	¼	¼	¼	¼	¼	¼	¼	¼	¼	¼	¼
Spacing—Inches	24	24	24	24	24	24	24	24	24	24	24	24
Chain												
Number and Style	14½ M.R.	126 C	126 C	14½ M.R.	126 C	126 C	14½ M.R.	126 C	126 C	14½ M.R.	126 C	126 C
Pitch—Inches	4.01	6	6	4.01	6	6	4.01	6	6	4.01	6	6
Attachments	T & M1 Sp.	T-Hy & M1 Sp.	T-Hy & M1 Sp.	T & M1 Sp.	T-Hy & M1 Sp.	T-Hy & M1 Sp.	T & M1 Sp.	T-Hy & M1 Sp.	T-Hy & M1 Sp.	T & M1 Sp.	T-Hy & M1 Sp.	T-Hy & M1 Sp.
Work Strength—Lbs	1600	3100	3100	1600	3100	3100	1600	3100	3100	1600	3100	3100
H. P. At Countershaft*	1.9	2.5	3.8	3.9	5.0	7.6	5.8	7.5	11.4	7.8	10.0	15.2
Head Shaft												
Diameter—Inches	1¹³⁄₁₆	2⁷⁄₁₆	2⁷⁄₁₆	2⁷⁄₁₆	2¹³⁄₁₆	2¹³⁄₁₆	2¹³⁄₁₆	3⁷⁄₁₆	3⁷⁄₁₆	3⁷⁄₁₆	3⁷⁄₁₆	3¹⁵⁄₁₆
Rev. per Min.	15½	16⅔	16⅔	15½	16⅔	16⅔	15½	16⅔	16⅔	15½	16⅔	16⅔
Size Sprocket—In.	24¾	23¾	23¾	24¾	23¾	23¾	24¾	23¾	23¾	24¾	23¾	23¾
Gear Diam—In.	29.83	29.83	29.83	29.83	29.83	40.12	29.83	40.12	40.12	40.12	40.12	41.24
Gear Pitch—In.	1¼	1¼	1¼	1¼	1¼	1½	1¼	1½	1½	1½	1½	1¾
Gear Face—In.	3	3	3	3	3	4	3	4	4	4	4	6
Countershaft												
Diameter—In.	1⁷⁄₁₆	1¹⁵⁄₁₆	1¹⁵⁄₁₆	1¹⁵⁄₁₆	2⁷⁄₁₆	2⁷⁄₁₆	2³⁄₁₆	2¹¹⁄₁₆	2¹¹⁄₁₆	2¹¹⁄₁₆	2¹¹⁄₁₆	2¹⁵⁄₁₆
Rev. per Min.	78	83	83	78	83	93	78	93	93	87	93	82
Pinion Diam.—In.	6.01	6.01	6.01	6.01	6.01	7.22	6.01	7.22	7.22	7.22	7.22	8.42
Pinion Face—In.	3¼	3¼	3¼	3¼	3¼	4½	3¼	4½	4½	4½	4½	6⅜
Foot Shaft												
Diameter—In.	1⁷⁄₁₆	1⁷⁄₁₆	1⁷⁄₁₆	1⁷⁄₁₆	1¹⁵⁄₁₆	1¹⁵⁄₁₆	1¹⁵⁄₁₆	2⁷⁄₁₆	2⁷⁄₁₆	2⁷⁄₁₆	2⁷⁄₁₆	2¹⁵⁄₁₆
Size Sprocket—In.	24¾	23¾	23¾	24¾	23¾	23¾	24¾	23¾	23¾	24¾	23¾	23¾
Trough												
Thickness or Gauge..	10	³⁄₁₆	³⁄₁₆	10	³⁄₁₆	³⁄₁₆	10	³⁄₁₆	³⁄₁₆	10	³⁄₁₆	³⁄₁₆
Approx. Shipping Wgt.—Lbs.												
Terminals, Complete	860	1300	1340	970	1500	1780	1160	1900	1920	1560	1900	2660
Chain and Flights per Ft. Ctrs.	24	42	48	24	42	48	24	42	48	24	42	48
Trough and Bar Trackage per Ft. Ctrs.	19½	24½	30	19½	24½	30	19½	24½	30	19½	24½	30

*For Maximum Centers for all inclinations and corresponding capacities.

"Weight of Terminals" given in table includes shafts, bearings, collars, sprockets and gears, with chain and flights half way around the sprockets.

Capacities are figured for 50 lb. material, troughs 80% level full, and uniformly loaded throughout the time period.

For Erection Dimensions of above Conveyors, see page 111.

General Dimensions of Jeffrey Standard Scraper Conveyors

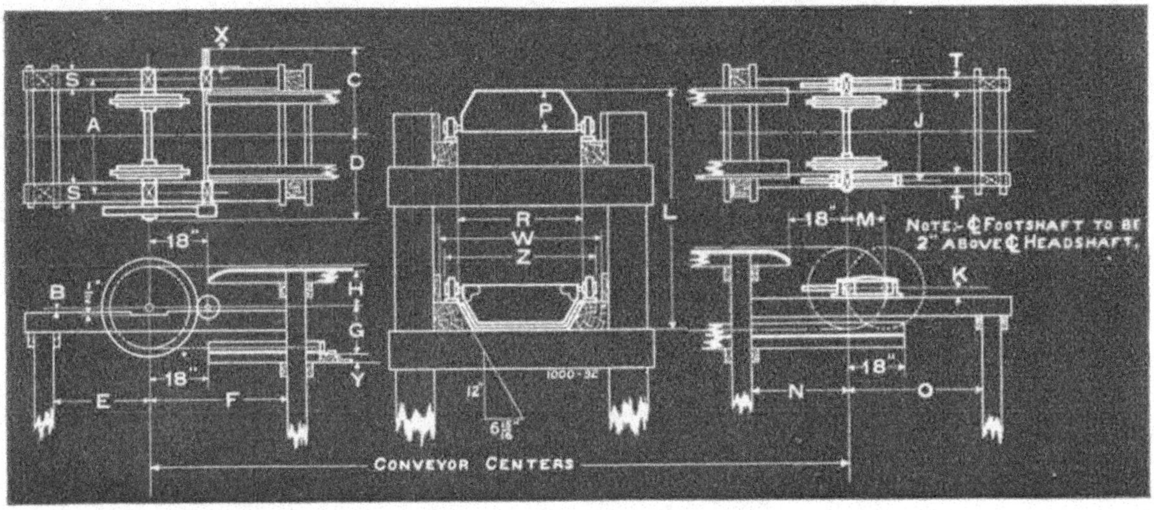

Using Double Strand Malleable Roller Chain with Steel Scrapers

Wood Supports—*For Steel Supports see page 113.* **Dimensions in Inches.**

Conveyor No.	A	B	C	D	E	F	G	H	J	K	L	M	N	O	P	R	S	T	W	X	Y	Z
2953	34¼	1 3/16	25⅜	24⅜	30	42	13⅝	11⅛	29	2¼	35⅞	11¾	30	42	6	19	4	3½	25	6	4 9/16	23½
2954	34¼	2⅝	26⅛	25⅞	30	42	13⅝	10⅛	33¼	2¼	34 11/16	11¾	30	42	6	20	6	3½	26⅞	6	3⅞	25⅜
2955	40¼	2⅝	29⅛	28⅞	32	42	13⅝	10⅛	39¼	2¼	38 11/16	11¾	30	44	8	26	6	3½	32⅞	6	5 15/16	31⅜
2956	33¼	2⅝	25⅝	25⅜	30	42	13⅝	11⅛	29	2¼	35⅞	11¾	30	42	6	19	6	3½	25	6	4 9/16	23½
2957	36	3⅛	27¾	27¾	30	42	13⅝	10⅛	31½	2¾	34 11/16	12	30	42	6	20	6	4	26⅞	6	3⅞	25⅜
2958	42	3⅛	30¾	30¾	32	48	13⅝	10⅛	37½	2¾	38 11/16	12	30	44	8	26	6	4	32⅞	6	5 15/16	31⅜
2959	35	3⅛	27¼	27¼	30	42	13⅝	11⅛	30½	2¾	35⅞	12	30	42	6	19	6	4	25	6	4 9/16	23½
2960	37¾	3½	30	29⅝	30	48	13⅝	10⅛	33¼	3⅛	34 11/16	15	32	46	6	20	8	6	26⅞	7	3⅞	25⅜
2961	43¾	3½	33	32⅝	32	48	13⅝	10⅛	39¼	3⅛	38 11/16	15	32	48	8	26	8	6	32⅞	7	5 15/16	31⅜
2962	36¾	3½	29½	29⅛	30	48	13⅝	11⅛	32¼	3⅛	35⅞	15	32	46	6	19	8	6	25	7	4 9/16	23½
2963	37¾	3½	30	29⅝	30	48	13⅝	10⅛	33¼	3⅛	34 11/16	15	32	46	6	20	8	6	26⅞	7	3⅞	25⅜
2964	46	4⅛	35½	35½	32	48	13⅝	10⅛	41	4	38 11/16	22¼	42	54	8	26	10	8	32⅞	8	5 15/16	31⅜

Specifications of Jeffrey Standard Scraper Conveyors using Double Strand Malleable Roller Chain with Steel Scrapers

Steel Supports—*For Wood Supports see page 110.*

Lgth. of Conveyor	0 to 50 ft. Centers			51 to 100 ft. Centers			101 to 150 ft. Centers			151 to 200 ft. Centers		
No. of Conveyor	3146	3147	3148	3149	3150	3151	3152	3153	3154	3155	3156	3157
Size of Material—In.												
Avge. size Material to be handled	6	6	8	6	6	8	6	6	8	6	6	8
Max. size; not to exceed 10% of whole	9	9	12	9	9	12	9	9	12	9	9	12
Capacity—In tons per hr.												
Horizontal	60	60	112	60	60	112	60	60	112	60	60	112
15° Incline	32	32	60	32	32	60	32	32	60	32	32	60
30° Incline	23	23	44	23	23	44	23	23	44	23	23	44
45° Incline	20	20	37	20	20	37	20	20	37	20	20	37
Size Scraper—In.												
Length	18	18	24	18	18	24	18	18	24	18	18	24
Depth	6	6	8	6	6	8	6	6	8	6	6	8
Thickness of Steel	¼	¼	¼	¼	¼	¼	¼	¼	¼	¼	¼	¼
Spacing—Inches	24	24	24	24	24	24	24	24	24	24	24	24
Chain												
Number and Style	14½ M.R.	126 C	126 C	14½ M.R.	126 C	126 C	14½ M.R.	126 C	126 C	14½ M.R.	126 C	126 C
Pitch—Inches	4.01	6	6	4.01	6	6	4.01	6	6	4.01	6	6
Attachments	T & M1 Sp.	T-Hy & M1 Sp.	T-Hy & M1 Sp.	T & M1 Sp.	T-Hy & M1 Sp.	T-Hy & M1 Sp.	T & M1 Sp.	T-Hy & M1 Sp.	T-Hy & M1 Sp.	T & M1 Sp.	T-Hy & M1 Sp.	T-Hy & M1 Sp.
Work Strength—Lbs.	1600	3100	3100	1600	3100	3100	1600	3100	3100	1600	3100	3100
H. P. At Counter-shaft*	1.9	2.5	3.8	3.9	5.0	7.6	5.8	7.5	11.4	7.8	10	15.2
Head Shaft												
Diameter—In.	1¹⁵⁄₁₆	2⁷⁄₁₆	2⁷⁄₁₆	2⁷⁄₁₆	2¹⁵⁄₁₆	2¹⁵⁄₁₆	2¹⁵⁄₁₆	3⁷⁄₁₆	3⁷⁄₁₆	3⁷⁄₁₆	3⁷⁄₁₆	3¹⁵⁄₁₆
Rev. per Min.	15½	16⅔	16⅔	15½	16⅔	16⅔	15½	16⅔	16⅔	15½	16⅔	16⅔
Size Sprocket—In.	24¾	23¾	23¾	24¾	23¾	23¾	24¾	23¾	23¾	24¾	23¾	23¾
Gear Diam.—In.	29.83	29.83	29.83	29.83	29.83	40.12	29.83	40.12	40.12	40.12	40.12	41.24
Gear Pitch—In.	1¼	1¼	1¼	1¼	1¼	1½	1¼	1½	1½	1½	1½	1¾
Gear Face—In.	3	3	3	3	3	4	3	4	4	4	4	6
Countershaft												
Diameter—In.	1⁷⁄₁₆	1¹⁵⁄₁₆	1¹⁵⁄₁₆	1¹⁵⁄₁₆	2⁷⁄₁₆	2⁷⁄₁₆	2⁷⁄₁₆	2¹¹⁄₁₆	2¹¹⁄₁₆	2¹¹⁄₁₆	2¹¹⁄₁₆	2¹⁵⁄₁₆
Rev. per Min.	78	83	83	78	83	93	78	93	93	87	93	82
Pinion Diam.—In.	6.01	6.01	6.01	6.01	6.01	7.22	6.01	7.22	7.22	7.22	7.22	8.42
Pinion Face—In.	3¼	3¼	3¼	3¼	3¼	4½	3¼	4½	4½	4½	4½	6⅜
Foot Shaft												
Diameter—In.	1⁷⁄₁₆	1⁷⁄₁₆	1⁷⁄₁₆	1⁷⁄₁₆	1¹⁵⁄₁₆	1¹⁵⁄₁₆	1¹⁵⁄₁₆	2⁷⁄₁₆	2⁷⁄₁₆	2⁷⁄₁₆	2⁷⁄₁₆	2¹⁵⁄₁₆
Size Sprocket—In.	24¾	23¾	23¾	24¾	23¾	23¾	24¾	23¾	23¾	24¾	23¾	23¾
Trough												
Thickness or Gauge	10	³⁄₁₆	³⁄₁₆	10	³⁄₁₆	³⁄₁₆	10	³⁄₁₆	³⁄₁₆	10	³⁄₁₆	³⁄₁₆
Approx. Shipping Wgt.—Lbs.												
Terminals, Complete	860	1300	1340	970	1500	1780	1160	1900	1920	1560	1900	2660
Chain and Flights per Ft. Ctrs.	24	42	48	24	42	48	24	42	48	24	42	48
Trough and Bar Trackage per Ft. Ctrs.	22	28	33½	22	28	33½	22	28	33½	22	28	33½

* For Maximum Centers for all inclinations and corresponding capacities.

"Weight of Terminals" given in table includes shafts, bearings, collars, sprockets and gears, with chain and flights half way around the sprockets.

Capacities are figured for 50 lb. material, troughs 80% level full, and uniformly loaded throughout the time period.

For Erection Dimensions of above Conveyors, see page 113.

General Dimensions of Jeffrey Standard Scraper Conveyors

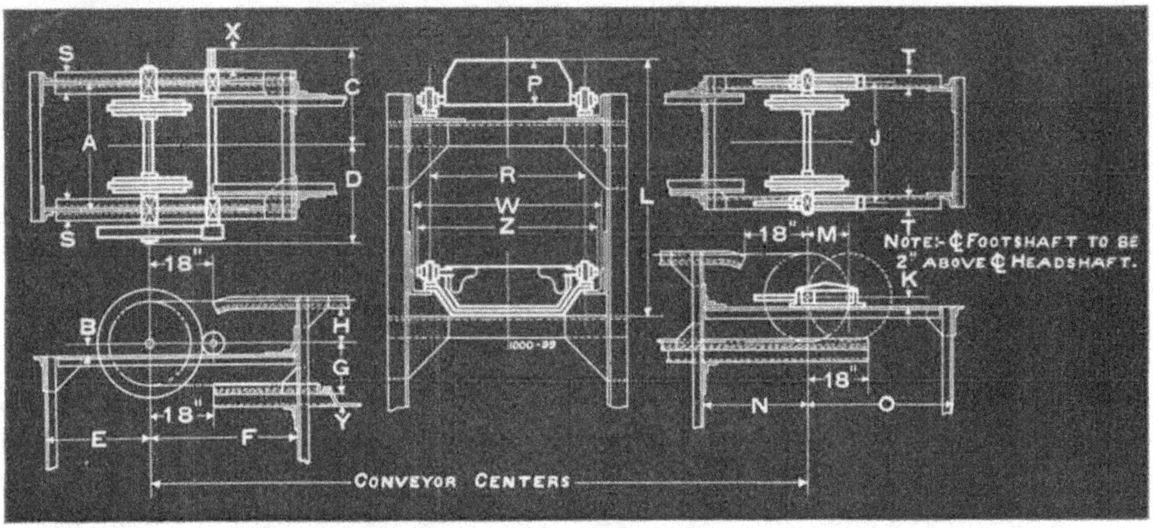

Using Double Strand Malleable Roller Chain with Steel Scrapers

Steel Supports—*For Wood Supports* *see page 111.* **Dimensions in Inches**

Conveyor No.	A	B	C	D	E	F	G	H	J	K	L	M	N	O	P	R	S	T	W	X	Y	Z
3146	34¼	1 11/16	25⅜	24⅜	30	42	14	10⅞	29	2¼	35⅞	11¾	30	42	6	21	*4	3½	25	6	4⅛	23½
3147	34¼	3⅛	26⅛	25⅞	30	42	14 1/16	9⅞	33¼	2¼	34 11/16	11¾	30	42	6	22	6¼	3½	26⅞	6	3½	25¼
3148	40¼	3⅛	29⅛	28⅞	32	42	14 1/16	9⅞	39¼	2¼	38 11/16	11¾	30	44	8	28	6¼	3½	32⅞	6	5½	31¼
3149	33¼	3⅛	25⅝	25⅜	30	42	14	10⅞	29	2¼	35⅞	11¾	30	42	6	21	6¼	3½	25	6	4⅛	23½
3150	36	3⅝	27¾	27¾	30	42	14 1/16	9⅞	31½	2¾	34 11/16	12	30	42	6	22	6¼	4	26⅞	6	3½	25¼
3151	42	3⅝	30¾	30¾	32	48	14 1/16	9⅞	37½	2¾	38 11/16	12	30	44	8	28	6¼	4	32⅞	6	5½	31¼
3152	35	3⅝	27¼	27¼	30	42	14	10⅞	30½	2¾	35⅞	12	30	42	6	21	6¼	4	25	6	4⅛	23½
3153	37¾	4	30	29⅝	30	48	14 1/16	9⅞	33¼	3⅛	34 11/16	15	32	46	6	22	7¼	4	26⅞	7	3½	25¼
3154	43¾	4	33	32⅝	32	48	14 1/16	9⅞	39¼	3⅛	38 11/16	15	32	48	8	28	7¼	4	32⅞	7	5½	31¼
3155	36¾	4	29½	29⅛	30	48	14	10⅞	32¼	3⅛	35⅞	15	32	46	6	21	7¼	4	25	7	4⅛	23½
3156	37¾	4	30	29⅝	30	48	14 1/16	9⅞	33¼	3⅛	34 11/16	15	32	46	6	22	7¼	4	26⅞	7	3½	25¼
3157	46	4⅝	35½	35½	32	48	14 1/16	9⅞	41	4	38 11/16	22¼	42	54	8	28	8¼	†6¼	32⅞	8	5½	31¼

* Use Single Angle for Head Bearing Supports.
† Use Double Angle for 2 15/16" Takeups.

Specifications of Jeffrey Standard Scraper Conveyors using Double Strand Vulcan Chain with Steel Scrapers

Wood Supports—*For Steel Supports see page 116.*

Length of Conveyor	0 to 50 ft. Centers		51 to 100 ft. Centers		101 to 150 ft. Centers		151 to 200 ft. Centers
No. of Conveyor	**2965**	**2966**	**2968**	**2969**	**2971**	**2972**	**2973**
Size of Material—In.							
Avge. size of Material to be handled	6	8	6	8	6	8	6
Max. size; not to exceed 10% of whole	9	12	9	12	9	12	9
Capacity—In tons per hour							
Horizontal	60	112	60	112	60	112	60
15° Incline	32	60	32	60	32	60	32
30° Incline	23	44	23	44	23	0	0
45° Incline	20	37	20	37	20	0	0
Size Scraper—In.							
Length	18	24	18	24	18	24	18
Depth	6	8	6	8	6	8	6
Thickness of Steel	1/4	1/4	1/4	1/4	1/4	1/4	1/4
Spacing—Inches	24	24	24	24	24	24	24
Chain							
Number and Style	526V	526V	526V	526V	526V	526V	526V
Pitch—Inches	6	6	6	6	6	6	6
Attachments	Bent Side Bar	Bent Side Bar	Bent Side Bar	Bent Side Bar	Bent Side Bar	Bent Side Bar	Bent Side Bar
Working Strength—Lbs	1640	1640	1640	1640	1640	1640	1640
H. P. at Countershaft*	2.6	4.0	5.2	8.0	7.8	12.1	10.4
Head Shaft							
Diameter—Inches	1 15/16	2 7/16	2 7/16	2 15/16	2 15/16	3 7/16	3 7/16
Rev. per Min.	16 1/6	16 2/3	16 2/3	16 2/3	16 2/3	16 2/3	16 2/3
Size Sprocket—Inches	23 1/2	23 1/2	23 1/2	23 1/2	23 1/2	23 1/2	23 1/2
Gear Diam.—Inches	29.83	29.83	29.83	40.12	40.12	40.12	40.12
Gear Pitch—Inches	1 1/4	1 1/4	1 1/4	1 1/2	1 1/2	1 1/2	1 1/2
Gear Face—Inches	3	3	3	4	4	4	4
Countershaft							
Diameter—Inches	1 7/16	1 15/16	1 15/16	2 7/16	2 7/16	2 11/16	2 11/16
Rev. per Min.	83	83	83	93	93	93	93
Pinion Diam.—Inches	6.01	6.01	6.01	7.22	7.22	7.22	7.22
Pinion Face—Inches	3 1/4	3 1/4	3 1/4	4 1/2	4 1/2	4 1/2	4 1/2
Foot Shaft							
Diameter—Inches	1 7/16	1 7/16	1 7/16	1 15/16	1 15/16	2 7/16	2 7/16
Size Sprocket—Inches	23 1/2	23 1/2	23 1/2	23 1/2	23 1/2	23 1/2	23 1/2
Trough							
Thickness—Inches	3/16	3/16	3/16	3/16	3/16	3/16	3/16
Approx. Shipping Wgt.—Lbs.							
Terminals, Complete	940	1090	1060	1510	1480	1660	1650
Chain and Flights per Ft. Ctrs.	32	38	32	38	32	38	32
Trough and Bar Trackage per Ft. Ctrs.	24 1/2	30	24 1/2	30	24 1/2	30	24 1/2

*For Maximum Centers for all inclinations and corresponding capacities.

"Weight of Terminals" given in table includes shafts, bearings, collars, sprockets and gears, with chain and flights half way around the sprockets.

Capacities are figured for 50 lb. material, troughs 80% level full, and uniformly loaded throughout the time period.

For Erection Dimensions of above Conveyors, see page 115.

General Dimensions of Jeffrey Standard Scraper Conveyors

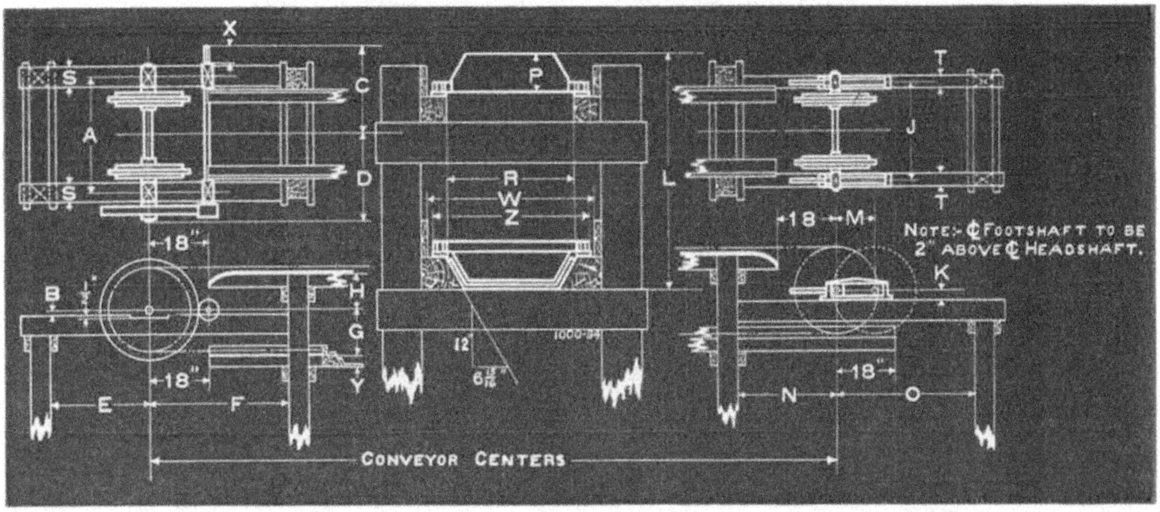

Using Double Strand Vulcan Chain with Steel Scrapers

Wood Supports—*For Steel Supports* *see page 117.*

Dimensions in Inches

Conveyor No.	A	B	C	D	E	F	G	H	J	K	L	M	N	O	P	R	S	T	W	X	Y	Z
2965	34¾	1 3/16	25⅝	24⅝	30	42	12¾	10¾	29½	2¼	34 7/16	11¾	30	42	6	19½	4	4	25⅛	6	4¾	23⅝
2966	39¾	2⅝	28⅞	28⅝	32	42	12¾	10¾	35½	2¼	38 7/16	11¾	30	44	8	25½	6	4	31⅛	6	6 11/16	29⅝
2968	33¾	2⅝	25⅞	25⅝	30	42	12¾	10¾	29½	2¼	34 7/16	11¾	30	42	6	19½	6	4	25⅛	6	4¾	23⅝
2969	41½	3⅛	30½	30½	32	48	12¾	10¾	37	2¾	38 7/16	12	30	44	8	25½	6	4	31⅛	6	6 11/16	29⅝
2971	35½	3⅛	27½	27½	30	48	12¾	10¾	31	2¾	34 7/16	12	30	42	6	19½	6	4	25⅛	6	4¾	23⅝
2972	43¼	3½	32¾	32⅜	32	48	12¾	10¾	38¾	3⅛	38 7/16	15	32	48	8	25½	8	6	31⅛	7	6 11/16	29⅝
2973	37¼	3½	29¾	29⅜	30	48	12¾	10¾	32¾	3⅛	34 7/16	15	32	46	6	19½	8	6	25⅛	7	4¾	23⅝

Specifications of Jeffrey Standard Scraper Conveyors using Double Strand Vulcan Chain with Steel Scrapers.

Steel Supports—*For Wood Supports see page 114.*

Length of Conveyor	0 to 50 ft. Centers		51 to 100 ft. Centers		101 to 150 ft. Centers		151 to 200 ft. Centers
No. of Conveyor	3158	3159	3161	3162	3164	3165	3166
Size of Material—In.							
Average size of Material to be handled............	6	8	6	8	6	8	6
Maximum size; not to exceed 10% of whole..	9	12	9	12	9	12	9
Capacity—In tons per hr.							
Horizontal..............	60	112	60	112	60	112	60
15° Incline..............	32	60	32	60	32	60	32
30° Incline..............	23	44	23	44	23	0	0
45° Incline..............	20	37	20	37	20	0	0
Size Scraper—In.							
Length.......................	18	24	18	24	18	24	18
Depth........................	6	8	6	8	6	8	6
Thickness of Steel..........	$\frac{1}{4}$	$\frac{1}{4}$	$\frac{1}{4}$	$\frac{1}{4}$	$\frac{1}{4}$	$\frac{1}{4}$	$\frac{1}{4}$
Spacing—Inches...........	24	24	24	24	24	24	24
Chain							
Number and Style.........	526V	526V	526V	526V	526V	526V	526V
Pitch—Inches..............	6	6	6	6	6	6	6
Attachments..............	Bent Side Bar	Bent Side Bar	Bent Side Bar	Bent Side Bar	Bent Side Bar	Bent Side Bar	Bent Side Bar
Working Strength –Lbs.	1640	1640	1640	1640	1640	1640	1640
H. P. at Countershaft*.	2.6	4.0	5.2	8.0	7.8	12.1	10.4
Head Shaft							
Diameter—Inches.........	$1\frac{13}{16}$	$2\frac{7}{16}$	$2\frac{7}{16}$	$2\frac{13}{16}$	$2\frac{13}{16}$	$3\frac{7}{16}$	$3\frac{7}{16}$
Rev. per Min................	$16\frac{2}{3}$	$16\frac{2}{3}$	$16\frac{2}{3}$	$16\frac{2}{3}$	$16\frac{2}{3}$	$16\frac{2}{3}$	$16\frac{2}{3}$
Size Sprocket—Inches ..	$23\frac{1}{2}$	$23\frac{1}{2}$	$23\frac{1}{2}$	$23\frac{1}{2}$	$23\frac{1}{2}$	$23\frac{1}{2}$	$23\frac{1}{2}$
Gear Diam.—Inches.....	29.83	29.83	29.83	40.12	40.12	40.12	40.12
Gear Pitch—Inches	$1\frac{1}{4}$	$1\frac{1}{4}$	$1\frac{1}{4}$	$1\frac{1}{2}$	$1\frac{1}{2}$	$1\frac{1}{2}$	$1\frac{1}{2}$
Gear Face—Inches......	3	3	3	4	4	4	4
Countershaft							
Diameter—Inches.........	$1\frac{7}{16}$	$1\frac{15}{16}$	$1\frac{15}{16}$	$2\frac{7}{16}$	$2\frac{7}{16}$	$2\frac{11}{16}$	$2\frac{11}{16}$
Rev. per Min................	83	83	83	93	93	93	93
Pinion Diam.—Inches .	6.01	6.01	6.01	7.22	7.22	7.22	7.22
Pinion Face—Inches......	$3\frac{1}{4}$	$3\frac{1}{4}$	$3\frac{1}{4}$	$4\frac{1}{2}$	$4\frac{1}{2}$	$4\frac{1}{2}$	$4\frac{1}{2}$
Foot Shaft							
Diameter—Inches.........	$1\frac{7}{16}$	$1\frac{7}{16}$	$1\frac{7}{16}$	$1\frac{15}{16}$	$1\frac{15}{16}$	$2\frac{7}{16}$	$2\frac{7}{16}$
Size Sprocket—Inches ..	$23\frac{1}{2}$	$23\frac{1}{2}$	$23\frac{1}{2}$	$23\frac{1}{2}$	$23\frac{1}{2}$	$23\frac{1}{2}$	$23\frac{1}{2}$
Trough							
Thickness—Inches	$\frac{3}{16}$	$\frac{3}{16}$	$\frac{3}{16}$	$\frac{3}{16}$	$\frac{3}{16}$	$\frac{3}{16}$	$\frac{3}{16}$
Approx.ShippingWgt.— Lbs.							
Terminals, Complete	940	1090	1060	1510	1480	1660	1650
Chain and Flights per Ft. Ctrs............	32	38	32	38	32	38	32
Trough and Bar Track- age per Ft. Ctrs.	$27\frac{1}{2}$	33	$27\frac{1}{2}$	33	$27\frac{1}{2}$	33	$27\frac{1}{2}$

* For Maximum Centers for all inclinations and corresponding capacities.
"Weight of Terminals" given in table includes shafts, bearings, collars, sprockets and gears, with chain and flights half way around the sprockets.
Capacities are figured for 50 lb. material, troughs 80% level full, and uniformly loaded throughout the time period.
For Erection Dimensions of above Conveyors, see page 117.

General Dimensions of Jeffrey Standard Scraper Conveyors

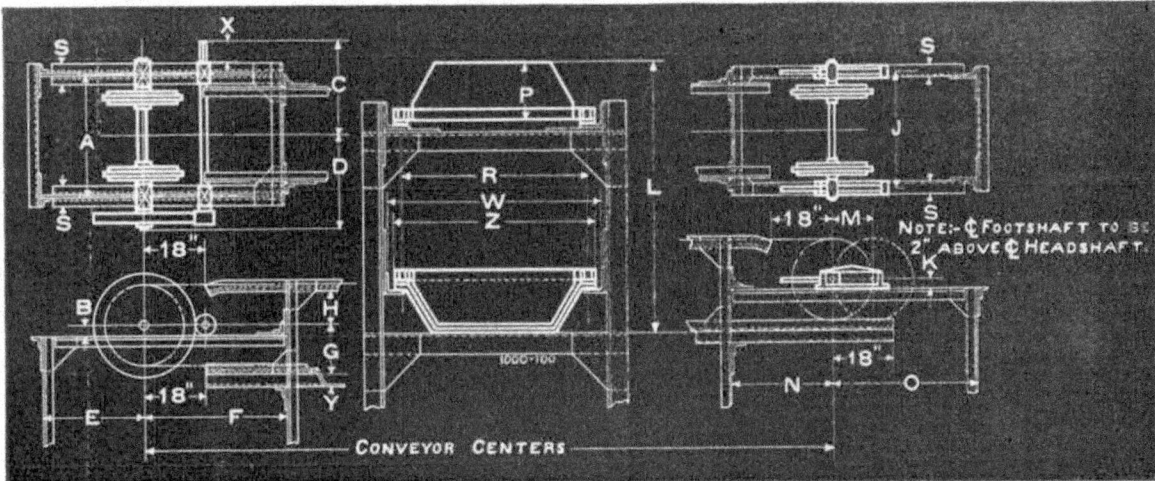

Using Double Strand Vulcan Chain with Steel Scrapers

Steel Supports—*For Wood Supports see page 115.* **Dimensions in Inches**

Conveyor No.	A	B	C	D	E	F	G	H	J	K	L	M	N	O	P	R	S	T	W	X	Y	Z
3158	34¾	1 11/16	25⅝	24⅝	30	42	13 3/16	10½	29½	2¼	34 7/16	11¾	30	42	6	21½	*4	3½	25½	6	4¼	23⅝
3159	39¾	3⅛	28⅞	28⅝	32	42	13 3/16	10½	35½	2¼	38 7/16	11¾	30	44	8	27½	6¼	3½	31½	6	6¼	29⅝
3161	33¾	3⅛	25⅞	25⅝	30	42	13 3/16	10½	29½	2¼	34 7/16	11¾	30	42	6	21½	6¼	3½	25½	6	4¼	23⅝
3162	41½	3⅝	30½	30½	32	48	13 3/16	10½	37	2¾	38 7/16	12	30	44	8	27½	6¼	4	31½	6	6¼	29⅝
3164	35½	3⅝	27½	27½	30	48	13 3/16	10½	31	2¾	34 7/16	12	30	42	6	21½	6¼	4	25½	6	4¼	23⅝
3165	43¼	4	32¾	32⅜	32	48	13 3/16	10½	38¾	3⅛	38 7/16	15	32	48	8	27½	7¼	4	31½	7	6¼	29⅝
3166	37¼	4	29¾	29⅜	30	48	13 3/16	10½	32¾	3⅛	34 7/16	15	32	46	6	21½	7¼	4	25½	7	4¼	23⅝

* Use Single Angle for Head Bearing Support.

Specifications of Jeffrey Standard Scraper Conveyors using Double Strand Steel Thimble Roller Chain with Steel Scrapers

Wood Supports—*For Steel Supports see page 120.*

Length of Conveyor	0 to 50 ft. Centers			51 to 100 ft. Centers			101 to 150 ft. Centers			151 to 200 ft. Centers		
No. of Conveyor	**2982**	**2983**	**2984**	**2985**	**2986**	**2987**	**2988**	**2989**	**2990**	**2991**	**2992**	**2993**
Size of Material—In.												
Avge. size of Material to be handled	8	10	12	8	10	12	8	10	12	8	10	12
Max. size; not to exceed 10% of whole	12	14	16	12	14	16	12	14	16	12	14	16
Capacity—In tons per hour												
Horizontal	92	167	238	92	167	238	92	167	238	92	167	238
15° Incline	50	90	129	50	90	129	50	90	129	50	90	129
30° Incline	36	65	93	36	65	93	36	65	93	36	65	93
45° Incline	31	56	80	31	56	80	31	56	80	31	56	0
Size Scraper—In.												
Length	24	30	36	24	30	36	24	30	36	24	30	36
Depth	8	10	12	8	10	12	8	10	12	8	10	12
Thickness of Steel	¼	¼	¼	¼	¼	¼	¼	¼	¼	¼	¼	¼
Spacing	24	36	36	24	36	36	24	36	36	24	36	36
Chain												
Number and Style	276 S.T.R	276 S.T.R	276 S.T.R	276 S.T.R	276 S.T.R	276 S.T.R	276 S.T.R	276 S.T.R	276 S.T.R	276 S.T.R	276 S.T.R	276 S.T.R
Pitch—Inches	12	12	12	12	12	12	12	12	12	12	12	12
Attachments	Flg'd Scraper	Flg'd Scraper	Flg'd Scraper	Flg'd Scraper	Flg'd Scraper	Flg'd Scraper	Flg'd Scraper	Flg'd Scraper	Flg'd Scraper	Flg'd Scraper	Flg'd Scraper	Flg'd Scraper
Work. Strength—Lbs.	5200	5200	5200	5200	5200	5200	5200	5200	5200	5200	5200	5200
H. P. at Countershaft*	3.8	5.5	7.4	7.6	11.0	14.7	11.4	16.5	22.1	15.2	22.0	29.3
Head Shaft												
Diameter—Inches	2 7/16	2 15/16	3 7/16	3 7/16	3 15/16	4 7/16	3 15/16	4 7/16	4 15/16	4 7/16	4 15/16	5 7/16
Rev. per Min.	16⅔	11	11	16⅔	11	11	16⅔	11	11	16⅔	11	11
Size Sprocket—Inches	24	35½	35½	24	35½	35½	24	35½	35½	24	35½	35½
Gear Diam.—Inches	29.83	40.12	40.12	40.12	41.24	41.24	40.12	48.41	41.24 C.S.	41.24	41.24 C.S.	41.24 C. S.
Gear Pitch—Inches	1¼	1½	1½	1½	1¾	1¾	1½	2	1¾	1¾	1¾	1¾
Gear Face—Inches	3	4	4	4	6	6	4	6	6	6	6	6
Countershaft												
Diameter—Inches	1 15/16	2 7/16	2 11/16	2 11/16	2 15/16	3 7/16	2 15/16	3 7/16	3 15/16	3 7/16	3 15/16	4 7/16
Rev. per Min.	83	62	62	93	54	54	93	56	54	82	54	54
Pinion Diam.—Inches	6.01	7.22	7.22	7.22	8.42	8.42	7.22	9.62	8.42 C.S.	8.42	8.42 C.S.	8.42 C.S.
Pinion Face—Inches	3¼	4½	4½	4½	6⅜	6⅜	4½	6½	6⅜	6⅜	6⅜	6⅜
Foot Shaft												
Diameter—Inches	1 7/16	1 15/16	2 7/16	2 7/16	2 7/16	2 15/16	2 7/16	2 15/16	2 15/16	2 15/16	2 15/16	2 15/16
Size Sprocket—In.	24	35½	35½	24	35½	35½	24	35½	35½	24	35½	35½
Trough												
Thickness—Inches	3/16	¼	¼	3/16	¼	¼	3/16	¼	¼	3/16	¼	¼
Approx. Shipping Wgt.—Lbs.												
Terminals, Complete	1620	2580	2580	2250	3230	3900	2440	4020	4200	3220	4120	4610
Chain and Flights per Ft. Ctrs.	72	72	80	72	72	80	72	72	80	72	72	80
Trough and Bar Trackage per Ft. Ctrs.	30	45	51½	30	45	51½	30	45	51½	30	45	51½

*For Maximum Centers for all inclinations and corresponding capacities.

"Weight of Terminals" given in table includes shafts, bearings, collars, sprockets and gears, with chain and flights half way around the sprockets.

Capacities are figured for 50 lb. material, troughs 80% level full, and uniformly loaded throughout the time period.

For Erection Dimensions of above Conveyors, see page 119.

General Dimensions of Jeffrey Standard Scraper Conveyors

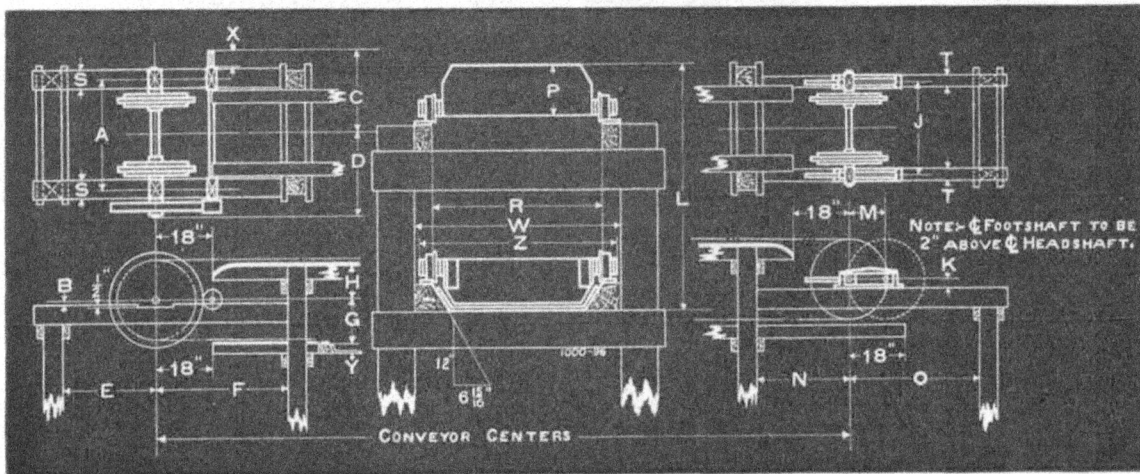

Using Double Strand Steel Thimble Roller Chain with Steel Scrapers

Wood Supports—*For Steel Supports see page 121.* **Dimensions in Inches**

Conveyor No.	A	B	C	D	E	F	G	H	J	K	L	M	N	O	P	R	S	T	W	X	Y	Z
2982	40⅛	2⅝	29 1/16	28 13/16	30¾	42	14¼	9¾	39⅜	2¼	38	11¾	30	42	8	26⅞	6	4	32¾	6	4¼	31¼
2983	47⅞	3⅛	33 11/16	33 11/16	40	48	20	15½	43⅜	2¾	53½	12	30	52	10	32⅞	6	4	38¾	6	6¼	37¼
2984	55⅝	3¼	38 13/16	38 9/16	40	48	20	15½	51⅛	3⅛	57½	15	32	56	12	38⅞	8	6	44¾	7	8¼	43¼
2985	43⅝	3¼	32 13/16	32 9/16	30¾	48	14¼	9¾	39⅜	3⅛	38	15	32	46	8	26⅞	8	6	32¾	7	4¼	31¼
2986	51⅞	4⅛	38 7/16	38 7/16	40	48	20	15½	45⅛	3⅛	53½	15	32	56	10	32⅞	8	6	38¾	8	6¼	37¼
2987	60⅛	4⅞	44 5/16	43 13/16	40	48	20	15½	52⅞	4	57½	22¼	44	62	12	38⅞	10	8	44¾	9	8¼	43¼
2988	45⅞	4⅛	35 7/16	35 3/16	30¾	48	14¼	9¾	39⅜	3⅛	38	15	32	46	8	26⅞	8	6	32¾	8	4¼	31¼
2989	54⅛	4⅞	41 5/16	40 13/16	40	56	20	15½	46⅞	4	53½	22¼	44	62	10	32⅞	10	8	38¾	9	6¼	37¼
2990	62⅜	5⅛	47 3/16	46 7/16	40	48	20	15½	52⅞	4	57½	22¼	44	62	12	38⅞	10	8	44¾	10	8¼	43¼
2991	48⅛	4⅞	38 5/16	37 13/16	30¾	48	14¼	9¾	40⅞	4	38	22¼	44	52	8	26⅞	10	8	32¾	9	4¼	31¼
2992	56⅜	5⅛	44 3/16	43 7/16	40	48	20	15½	46⅞	4	53½	22¼	44	62	10	32⅞	10	8	38¾	10	6¼	37¼
2993	64⅝	5¾	50 1/16	49 1/16	40	48	20	15½	52⅞	4	57½	22¼	44	62	12	38⅞	12	8	44¾	11	8¼	43¼

Specifications of Jeffrey Standard Scraper Conveyors using Double Strand Steel Thimble Roller Chain with Steel Scrapers

Steel Supports—*For Wood Supports see page 118*

Length of Conveyor	0 to 50 ft. Centers			51 to 100 ft. Centers			101 to 150 ft. Centers			151 to 200 ft. Centers		
No. of Conveyor	3175	3176	3177	3178	3179	3180	3181	3182	3183	3184	3185	3186
Size of Material—In.												
Avge. size of Material to be handled	8	10	12	8	10	12	8	10	12	8	10	12
Max. size; not to exceed 10% of whole	12	14	16	12	14	16	12	14	16	12	14	16
Capacity—In tons per hour												
Horizontal	92	167	238	92	167	238	92	167	238	92	167	238
15° Incline	50	90	129	50	90	129	50	90	129	50	90	129
30° Incline	36	65	93	36	65	93	36	65	93	36	65	93
45° Incline	31	56	80	31	56	80	31	56	80	31	56	0
Size Scraper—In.												
Length	24	30	36	24	30	36	24	30	36	24	30	36
Depth	8	10	12	8	10	12	8	10	12	8	10	12
Thickness of Steel	$\frac{1}{4}$	$\frac{1}{4}$	$\frac{1}{4}$	$\frac{1}{4}$	$\frac{1}{4}$	$\frac{1}{4}$	$\frac{1}{4}$	$\frac{1}{4}$	$\frac{1}{4}$	$\frac{1}{4}$	$\frac{1}{4}$	$\frac{1}{4}$
Spacing	24	36	36	24	36	36	24	36	36	24	36	36
Chain												
Number and Style	276 S.T.R	276 S.T.R	276 S.T.R	276 S.T.R	276 S.T.R	276 S.T.R	276 S.T.R	276 S.T.R	276 S.T.R	276 S.T.R	276 S.T.R	276 S.T.R
Pitch—Inches	12	12	12	12	12	12	12	12	12	12	12	12
Attachments	Flg'd Scraper	Flg'd Scraper	Flg'd Scraper	Flg'd Scraper	Flg'd Scraper	Flg'd Scraper	Flg'd Scraper	Flg'd Scraper	Flg'd Scraper	Flg'd Scraper	Flg'd Scraper	Flg'd Scraper
Work. Strength—Lbs.	5200	5200	5200	5200	5200	5200	5200	5200	5200	5200	5200	5200
H. P. at Countershaft*	3.8	5.5	7.4	7.6	11.0	14.7	11.4	16.5	22.1	15.2	22.0	29.3
Head Shaft												
Diameter—Inches	$2\frac{5}{16}$	$2\frac{15}{16}$	$3\frac{7}{16}$	$3\frac{7}{16}$	$3\frac{15}{16}$	$4\frac{7}{16}$	$3\frac{15}{16}$	$4\frac{7}{16}$	$4\frac{15}{16}$	$4\frac{7}{16}$	$4\frac{15}{16}$	$5\frac{7}{16}$
Rev. per Min.	$16\frac{2}{3}$	11	11	$16\frac{2}{3}$	11	11	$16\frac{2}{3}$	11	11	$16\frac{2}{3}$	11	11
Size Sprocket—In.	24	$35\frac{1}{2}$	$35\frac{1}{2}$	24	$35\frac{1}{2}$	$35\frac{1}{2}$	24	$35\frac{1}{2}$	$35\frac{1}{2}$	24	$35\frac{1}{2}$	$35\frac{1}{2}$
Gear Diam.—In.	29.83	40.12	40.12	40.12	41.24	41.24	40.12	48.41	41.24 C.S.	41.24	41.24 C.S.	41.24 C.S.
Gear Pitch—In.	$1\frac{1}{4}$	$1\frac{1}{2}$	$1\frac{1}{2}$	$1\frac{1}{2}$	$1\frac{3}{4}$	$1\frac{3}{4}$	$1\frac{1}{2}$	2	$1\frac{3}{4}$	$1\frac{3}{4}$	$1\frac{3}{4}$	$1\frac{3}{4}$
Gear Face—In.	3	4	4	4	6	6	4	6	6	6	6	6
Countershaft												
Diameter—In.	$1\frac{15}{16}$	$2\frac{7}{16}$	$2\frac{11}{16}$	$2\frac{11}{16}$	$2\frac{15}{16}$	$3\frac{7}{16}$	$2\frac{15}{16}$	$3\frac{7}{16}$	$3\frac{15}{16}$	$3\frac{7}{16}$	$3\frac{15}{16}$	$4\frac{7}{16}$
Rev. per Min.	83	62	62	93	54	54	93	56	54	82	54	54
Pinion Diam.—In.	6.01	7.22	7.22	7.22	8.42	8.42	7.22	9.62	8.42 C.S.	8.42	8.42 C.S.	8.42 C.S.
Pinion Face—In.	$3\frac{1}{4}$	$4\frac{1}{2}$	$4\frac{1}{2}$	$4\frac{1}{2}$	$6\frac{3}{8}$	$6\frac{3}{8}$	$4\frac{1}{2}$	$6\frac{1}{2}$	$6\frac{3}{8}$	$6\frac{3}{8}$	$6\frac{3}{8}$	$6\frac{3}{8}$
Foot Shaft												
Diameter—In.	$1\frac{7}{16}$	$1\frac{15}{16}$	$2\frac{7}{16}$	$2\frac{7}{16}$	$2\frac{7}{16}$	$2\frac{15}{16}$	$2\frac{7}{16}$	$2\frac{15}{16}$	$2\frac{15}{16}$	$2\frac{15}{16}$	$2\frac{15}{16}$	$2\frac{15}{16}$
Size Sprocket—In.	24	$35\frac{1}{2}$	$35\frac{1}{2}$	24	$35\frac{1}{2}$	$35\frac{1}{2}$	24	$35\frac{1}{2}$	$35\frac{1}{2}$	24	$35\frac{1}{2}$	$35\frac{1}{2}$
Trough												
Thickness	$\frac{3}{16}$	$\frac{1}{4}$	$\frac{1}{4}$	$\frac{3}{16}$	$\frac{1}{4}$	$\frac{1}{4}$	$\frac{3}{16}$	$\frac{1}{4}$	$\frac{1}{4}$	$\frac{3}{16}$	$\frac{1}{4}$	$\frac{1}{4}$
Approx. Shipping Wgt.—Lbs.												
Terminals, Complete	1620	2580	2850	2250	3230	3900	2440	4020	4200	3220	4120	4610
Chain and Flights per Ft. Ctrs.	72	72	80	72	72	80	72	72	80	72	72	80
Trough and Bar Trackage per Ft. Ctrs.	33	49	56	33	49	56	33	49	56	33	49	56

* For Maximum Centers for all inclinations and corresponding capacities.

"Weight of Terminals" given in table includes shafts, bearings, collars, sprockets and gears, with chain and flights half way around the sprockets.

Capacities are figured for 50 lb. material, troughs 80% level full, and uniformly loaded throughout the time period.

For Erection Dimensions of above Conveyors, see page 121.

General Dimensions of Jeffrey Standard Scraper Conveyors

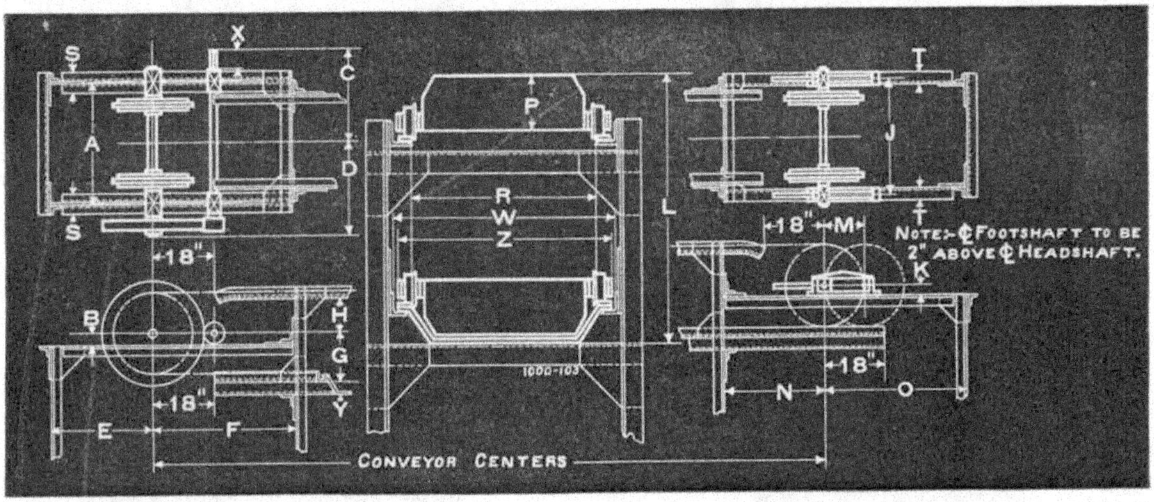

Using Double Strand Steel Thimble Roller Chain with Steel Scrapers

Steel Supports—*For Wood Supports see page 119.* **Dimensions in Inches**

Conveyor No.	A	B	C	D	E	F	G	H	J	K	L	M	N	O	P	R	S	T	W	X	Y	Z
3175	40 1/8	3 1/8	29 1/16	28 13/16	30 3/4	42	14 11/16	9 1/2	39 1/8	2 1/4	38	11 3/4	30	42	8	26 7/8	6 1/4	4	32 3/4	6	4 1/2	31 1/4
3176	47 7/8	3 5/8	33 11/16	33 11/16	40	48	20 1/2	15 1/4	43 3/8	2 3/4	53 1/2	12	30	52	10	32 7/8	6 1/4	4	38 3/4	6	6 1/2	37 1/4
3177	55 5/8	4	38 15/16	38 9/16	40	48	20 1/2	15 1/4	51 1/8	3 1/8	57 1/2	15	32	56	12	38 7/8	7 1/4	4	44 3/4	7	8 1/2	43 1/4
3178	43 5/8	4	32 15/16	32 9/16	30 3/4	48	14 11/16	9 1/2	39 1/8	3 1/8	38	15	32	46	8	26 7/8	7 1/4	4	32 3/4	7	4 1/2	31 1/4
3179	51 7/8	4 5/8	38 7/16	38 7/16	40	48	20 1/2	15 1/4	45 1/8	3 1/8	53 1/2	15	32	56	10	32 3/4	8 1/4	4	38 3/4	8	6 1/2	37 1/4
3180	60 1/8	5 3/8	44 5/16	43 13/16	40	48	20 1/2	15 1/4	52 7/8	4	57 1/2	22 1/4	44	62	12	38 7/8	10 1/4	*6 1/4	44 3/4	9	8 1/2	43 1/4
3181	45 7/8	4 5/8	35 7/16	35 3/16	30 3/4	48	14 11/16	9 1/2	39 1/8	3 1/8	38	15	32	46	8	26 7/8	8 1/4	4	32 3/4	8	4 1/2	31 1/4
3182	54 1/8	5 3/8	41 5/16	40 13/16	40	56	20 1/2	15 1/4	46 7/8	4	53 1/2	22 1/4	44	62	10	32 7/8	10 1/4	*6 1/4	38 3/4	9	6 1/2	37 1/4
3183	62 3/8	5 5/8	47 3/16	46 7/16	40	48	20 1/2	15 1/4	52 7/8	4	57 1/2	22 1/4	44	62	12	38 7/8	10 1/4	*6 1/4	44 3/4	10	8 1/2	43 1/4
3184	48 1/8	5 3/8	38 5/16	37 13/16	30 3/4	48	14 11/16	9 1/2	40 7/8	4	38	22 1/4	44	52	8	26 7/8	10 1/4	*6 1/4	32 3/4	9	4 1/2	31 1/4
3185	56 3/8	5 5/8	44 3/16	43 7/16	40	48	20 1/2	15 1/4	46 7/8	4	53 1/2	22 1/4	44	62	10	32 7/8	10 1/4	*6 1/4	38 3/4	10	6 1/2	37 1/4
3186	64 5/8	6 1/4	50 1/16	49 1/16	40	48	20 1/2	15 1/4	52 7/8	4	57 1/2	22 1/4	44	62	12	38 7/8	12 1/4	*6 1/4	44 3/4	11	8 1/2	43 1/4

* Use Double Angles for 2 15/16" Takeups.

Jeffrey Reliance Drag Chain Ashes Conveyor removing Ashes from in front of boilers. It operates at a very slow speed, with a small amount of power and thus easily handles the ashes of plants having comparatively few boilers.

MANY power plant engineers prefer to have separate conveyors for handling Ashes and Coal, as a shortage of coal supply or delayed delivery, under certain operating conditions, often necessitates the handling of coal at a time when the ash hoppers under the stokers require emptying.

Illustration at right shows Jeffrey Reliance Drag Chain
Conveyor delivering ashes from Boiler Room
to railroad cars.

Jeffrey Reliance Drag Chain

Here the Jeffrey Drag Chain Conveyor operates just below the floor level in an extra heavy cast iron trough set in a cement trench. This arrangement makes it easy for the man tending the boilers to watch the ash conveyor also. This Conveyor may also be located in the basement and fed by raking from ash pits.

Handling Ashes with a Conveyor adapted for Gritty Material

Where conditions permit the Drag Chain Equipment to be installed for handling ashes, the cost of maintenance is greatly reduced. .

This cost reduction is possible for two reasons: (1st) The Drag Chain itself is a low priced chain which means a low cost for repairs. (2nd) The actual wear on the chain is greatly reduced because a film of finer material ordinarily acts as a cushion or wearing surface between the chain and the trough.

The long wearing surface in the Jeffrey Drag Chain for its pins makes it ideally fitted to Conveyors for the handling of Ashes, during the 3 to 5 hour continuous daily service required of it in many power plants.

Jeffrey Reliance Drag Chains are furnished in two sizes of similar construction. The Number 102 Chain has a 5″ pitch, 9⅞″ overall width and 4200 pounds working strength. The Number 1156 Chain has a 6″ pitch, 9¾″ overall width, and working strength 5000 pounds.

These Drag Chain Conveyors operate in a cast iron trough and handle a capacity of 20 tons per hour, at a chain speed of 50 ft. per minute. For any lesser capacities change the chain speed in direct proportion.

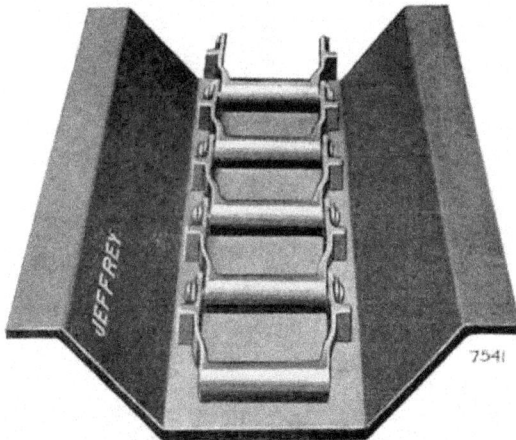

The Cast Iron Trough used in connection with Jeffrey Drag Chain Conveyors is made extra thick.

Specify Conveyor by Number given in Table.

Specifications of Reliance Drag Chain Conveyor

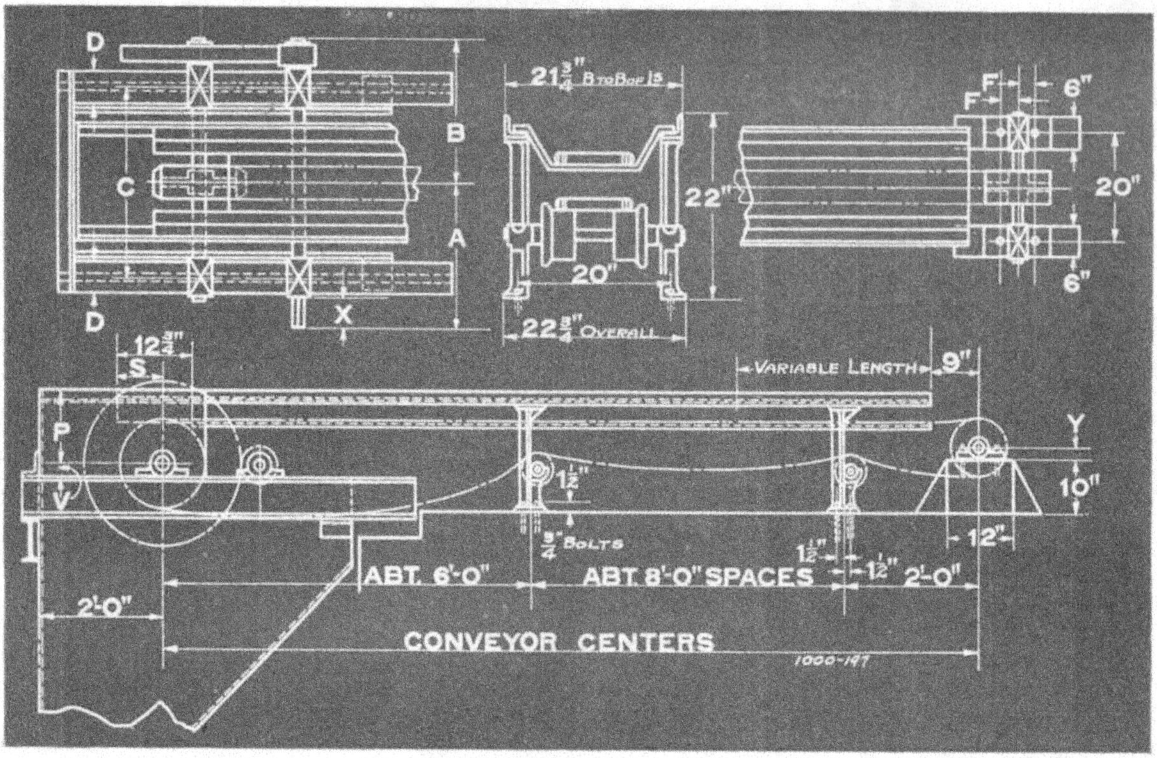

Length of Conveyor	0 to 50 ft. Centers		51 to 100 ft. Centers		101 to 150 ft. Centers		151 to 200 ft. Centers	
No. of Conveyor	3191	3192	3193	3194	3195	3196	3197	3198
Capacity—In tons per hour	20	20	20	20	20	20	20	20
No. of Chain	102	1156	102	1156	102	1156	102	1156
Pitch—Inches	5	6	5	6	5	6	5	6
Width—Inches	9⅞	9¾	9⅞	9¾	9⅞	9¾	9⅞	9¾
Pin Diameter—Inches	⅝	¾	⅝	¾	⅝	¾	⅝	¾
H. P. At Countershaft for Max. Centers	1.6	1.6	3.2	3.2	4.8	4.8	6.4	6.4
Head Shaft Diam.—Inches	2 7/16	2 7/16	2 11/16	2 11/16	2 11/16	2 11/16	3 7/16	3 7/16
Head Shaft R. P. M.	12	11	12	11	12	11	12	11
Drive Sprocket Diam.—Inches	16¼	17½	16¼	17½	16¼	17½	16¼	17½
Drive Gear Diam.—Inches	29.83	29.83	29.83	29.83	35.82	35.82	35.82	35.82
Drive Gear Pitch—Inches	1¼	1¼	1¼	1¼	1½	1½	1½	1½
Countershaft Diam.—Inches	1 15/16	1 15/16	2 7/16	2 7/16	2 7/16	2 7/16	2 11/16	2 11/16
Countershaft R. P. M.	60	55	60	55	60	55	60	55
Pinion Diam.—Inches	6.01	6.01	6.01	6.01	7.22	7.22	7.22	7.22
Pinion Face—Inches	3¼	3¼	3¼	3¼	4½	4½	4½	4½
Foot Shaft Diam.—Inches	1 15/16	1 15/16	1 15/16	1 15/16	2 7/16	2 7/16	2 7/16	2 7/16
C. I. Trough Pat. No.	29349	29349	29349	29349	29349	29349	29349	29349
C. I. Trough Wt. Per Ft.	36	36	36	36	36	36	36	36
Approx. Shipping Wgt.—Lbs.								
* Terminals, Complete	620	640	740	760	940	960	1080	1100
** Per Foot Centers	75	85	75	85	75	85	75	85

* Weight of Terminals—Include Head and Foot Shafts complete with Chain on End Sprockets.

**Weight per Foot Centers—Include Trough, Chains and Stands spaced 8 feet between centers with Rollers, Bearings and Shafts, also angles for Trough.

General Dimensions of Reliance Drag Chain Conveyors

0 to 50 ft. Centers

No.	A	B	C	D	F	P	S	V	X	Y
3191	26¼	26	34½	6	2 11/16	13¼	6¼	2 1/16	6	1 11/16
3192	26¼	26	34½	6	2 11/16	13¼	6	2 1/16	6	1 11/16

51 to 100 ft. Centers

No.	A	B	C	D	F	P	S	V	X	Y
3193	27¾	27¾	36	6	2 11/16	13¼	6¼	2 7/16	6	1 11/16
3194	27¾	27¾	36	6	2 11/16	13¼	6	2 7/16	6	1 11/16

101 to 150 ft. Centers

No.	A	B	C	D	F	P	S	V	X	Y
3195	28⅜	28⅜	37¼	7	3½	13¼	6¼	2 7/16	6	2 1/16
3196	28⅜	28⅜	37¼	7	3½	13¼	6	2 7/16	6	2 1/16

151 to 200 ft. Centers

No.	A	B	C	D	F	P	S	V	X	Y
3197	30⅞	30⅜	39¾	7	3½	13¼	6¼	2¾	7	2 1/16
3198	30⅞	30⅜	39¾	7	3½	13¼	6	2¾	7	2 1/16

Belt
Conveyors

Section
7

Jeffrey Three-Pulley Type Belt Conveyor distributing coal into bunkers by means of Traveling Tripper. Note the large amount of material handled by this Belt.

Another Three-Pulley Belt Conveyor handling coal in a large Boiler House.

Self-Propelled Reversing Belt Tripper, which automatically distributes as it discharges on each side of belt.

Here a Jeffrey Belt Conveyor, partly inclined and partly horizontal handles crushed coal from under track hopper to bunkers in Boiler House.

THE Belt Conveyor has one of its most economical applications in the distribution of coal into long bins or over long storage spaces. This is especially true where local conditions permit one belt to carry up a long incline and into bins.

The Shuttle Conveyor, as its name implies, shuttles or moves along a track over a bin or storage pile as illustrated below. It is used with a cross conveyor or an elevator operating at right angles to the length of the storage space and is so located as to permit the discharge of material onto the Shuttle Conveyor at about the center of the storage. Thus where local conditions permit, the Shuttle Belt is a minimum of machinery for filling a large storage space.

Cross section of Belt Conveyor showing the Jeffrey Standard 5-pulley troughing idler and side hanging return idler. Note that face of troughing pulleys conforms to natural troughing effect of belt for maximum carrying capacity.

Jeffrey Self-Contained Shuttle Belt Conveyor

General Rules and Ready References for Belt Conveyors
A Handy Guide for the Busy Engineer

THE following general rules, which embody the best in Belt Conveyor practice, have been collected out of the various pages of this section, for the convenience of the engineer who has not the time to go further into details.

Capacity in Tons Per Hour of materials weighing 50 pounds per cubic foot and carried over Three or Five Pulley Troughing Carriers is approximately 8% of the square of the number of inches in the width of the belt, for each 100 feet of belt speed per minute.

Maximum Speeds of Belt for loose materials equal approximately an initial speed of 250 feet per minute for a ten inch belt, plus 10 feet per minute for each additional inch of belt width. Slow speed 150 feet per minute. See page 130.

Horse Power Required: Approximately 2% of the number of tons per hour carried for each 100 feet of horizontal belt length and 1% additional for each 10 feet rise of incline. Increase this Horse Power at the belt 5% for each driving reduction thru chain, belting or cut gears and 10% for each reduction through rough gears to obtain the final Horse Power at Line Shaft, Motor or Engine. See page 131.

Maximum Inclines: Belts are ordinarily installed to carry horizontally although most materials may be readily carried in a troughed belt at 18 degrees—20 degrees to the horizontal, many others as high as 21 degrees—23 degrees and some few at 25 degrees. Recommendations covering angles over 20 degrees will be made upon statement of material to be handled, moisture in same, and nature of installation. See page 132.

Belt Values: Proper Flexibility in Belts for Troughing Carriers is one ply for each 4 to 5 inches of belt width, with 12 inch, 3 ply as a minimum and 48 inch, 8 ply as a maximum in ordinary service. See page 142.

Ultimate Strength of the average Rubber Belt is 360 pounds per inch width of each ply, with a Safe Working Tension of 30 pounds per inch width of each ply. Factor of Safety 12. The pull required to move a belt over its carriers upon the level is approximately 20% of weight of belt plus 10% of weight of load upon the belt.

Terminal Pulleys: The diameter of drive pulleys in ordinary good practice is 5 times the number of belt plies with the diameter of all other pulleys taking 180 degrees wrap 4 times the number of plies.

Spacing of Idlers: For Conveyors handling materials weighing up to 100 pounds per cubic foot and operating at full capacity, Troughing Carriers should be spaced as shown by "A" in table below.

End Troughing Carriers should be placed close to head and foot pulleys, as given by "B" in table, in order to prevent material from spreading to edges of belt.

Belt Width	A	B	Belt Width	A	B
14″ to 16″	5′-0″	2′-6″	24″ to 30″	4′-0″	3′-0″
18″ to 20″	4′-6″	3′-0″	36″ to 48″	3′-6″	3′-0″

At a loading chute, space the Carriers 24″ to 30″ apart under the chute with the first carrier 6″ back of the loading point, but never directly under the loading point, where material first touches the belt, as this somewhat increases the wear of the belt and often causes that carrier to be broken.

Loaders: For suggestions on the best methods of loading, see page 133.

Unloaders: Much of the life of a belt conveyor depends upon its Unloaders, see pages 139 and 140.

Cleaning Brushes: Cleaning brush speeds are ordinarily 800 to 1000 feet per minute at the surface for dry dusty materials; 1000 to 1200 for damp materials; 1200 to 1500 for wet, sticky materials.

Index and Capacity for Standard Belt Conveyors

Weight of Material to be Handled	Average Size Pieces	Maximum Size Pieces	Capacity per Hour — Tons	Cu. Ft.	Cu. Yds.	Bushels	Width of Belt	Speed in Feet per Minute	0 to 100 Conv. No.	0–100 Spec	0–100 Dim	101 to 200 Conv. No.	101–200 Spec	101–200 Dim	201 to 300 Conv. No.	201–300 Spec	201–300 Dim	301 to 400 Conv. No.	301–400 Spec	301–400 Dim	401 to 500 Conv. No.	401–500 Spec	401–500 Dim	501 to 600 Conv. No.	501–600 Spec	501–600 Dim	Carriers
50 Lb.	2	3	36	1456	54	1174	14	225	301	144	150	313	145	150	325	146	150	337	147	151	349	148	151	361	149	151	151 (3 Pulley)
	2½	4	48	1915	71	1539	16	225	302	144	150	314	145	150	326	146	150	338	147	151	350	148	151	362	149	151	151 (3 Pulley)
	3	5	70	2800	104	2250	18	260	303	144	150	315	145	150	327	146	150	339	147	151	351	148	151	363	149	151	151 (3 Pulley)
	3½	6	86	3458	128	2779	20	260	304	144	150	316	145	150	328	146	150	340	147	151	352	148	151	364	149	151	151 (3 Pulley)
	4½	8	143	5746	213	4617	24	300	306	144	150	318	145	150	330	146	150	342	147	151	354	148	151	366	149	151	151 (5 Pulley)
	6	11	254	10175	377	8175	30	340	309	144	150	321	145	150	333	146	150	345	147	151	357	148	151	369	149	151	151 (5 Pulley)
	7½	14	404	16160	599	12085	36	375	312	144	150	324	145	150	336	146	150	348	147	151	360	148	151	372	149	151	151 (5 Pulley)
	9	17	601	24048	892	19323	42	410	421	144	150	426	145	150	428	146	150	430	147	151	432	148	151	434	149	151	151 (5 Pulley)
	10½	20	862	34474	1278	27702	48	450	422	144	150	427	145	150	429	146	150	431	147	151	433	148	151	435	149	151	151 (5 Pulley)

Proper Width of Belt for Size of Material

WIDTH of Belt for Size of Material is to be determined from the following table whether the capacities be large or small. The size of material listed for the various belts may be any size smaller but not larger than the average cube sizes given in the table.

Width of Belt for a given capacity should be determined by using 95% of the capacities given in Tables on following pages but in no case should it be less than the minimum width for the size of material handled as given above.

Capacities given in the Tables on the pages following are based upon a continuous uniform flow of material to the conveyor throughout the unit of time specified, and in choosing a width and speed of belt, care must be taken that the quantity of material to be carried can be fed to the belt under the operating conditions imposed.

Belt Width	14"	16"	18"	20"	24"	30"	36"	42"	48"
Size of Material — Uniform Size or 70% to 80% of Unsized material with	2"	2½"	3"	3½"	4½"	6"	7½"	9"	10½"
Max. Unsized pieces not over 10% of all.	3"	4"	5"	6"	8"	11"	14"	17"	20"

Capacities of Belt Conveyors using Three Pulley Carriers, page 134

Width of Belt Inches	Cross Section of Load in sq. ft.	Cu. Ft. per Hr. at 100 F. P. M.	Cu. Yds. per Hr. at 100 F. P. M.	Bushels per Hr. at 100 F. P. M.	Tons per Hour at 100 Feet per Min.				
					Weight of material in Lbs. per Cubic Foot				
					25	50	75	100	125
14	.114	686	25.4	551	8.6	17.2	25.8	34.4	43.0
16	.149	896	33.2	720	11.2	22.4	33.6	44.8	56.0
18	.189	1134	42.0	911	14.2	28.4	42.6	56.8	71.0
20	.233	1400	51.9	1125	17.5	35.0	52.5	70.0	87.5
42	1.029	6174	229.0	4961	77.2	154.4	231.6	308.8	386.0
48	1.344	8064	299.0	6480	100.8	201.6	302.4	403.2	504.0

F. P. M. = Feet Per Minute.

Above Table Based on Cross Section shown

Formula—Area of Section in Square Feet = $.000583W^2$.

Cubic Feet per Hour at 100 feet per minute speed of Conveyor = $3.5 W^2$.

Where W = Width of Belt in inches.

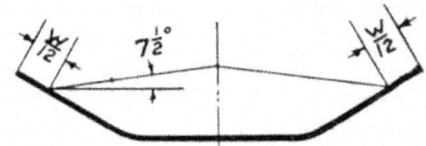

Outline of Belt and Bulk Loading for Three Pulley Carrier

Capacities of Belt Conveyors using Five Pulley Carriers, page 135

Width of Belt Inches	Cross Section of Load in Sq. Ft.	Cu. Ft. per Hr. at 100 F. P. M.	Cu. Yds. per Hr. at 100 F. P. M.	Bushels per Hr. at 100 F. P. M.	Tons Per Hour at 100 Feet per Minute				
					Weight of Material in Lbs. per Cubic Foot				
					25	50	75	100	125
24	.336	2016	74.7	1620	25.2	50.4	75.6	100.8	126.0
30	.525	3150	116.7	2531	39.4	78.8	118.2	157.6	197.0
36	.756	4536	168.0	3645	56.7	113.4	170.1	226.8	283.5
42	1.029	6174	229.0	4961	77.2	154.4	231.6	308.8	386.0
48	1.344	8064	299.0	6480	100.8	201.6	302.4	403.2	504.0

F. P. M. = Feet per Minute.

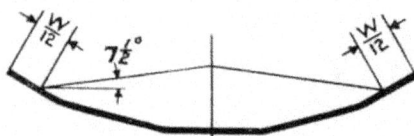

Outline of Belt and Bulk Loading for Five Pulley Carrier

Above Table Based on Cross Section Shown

Formula—Area of Section in Square Feet = $.000583 W^2$.

Cubic Feet per Hour at 100 feet per minute speed of Conveyor = $3.5 W^2$.

Where W = Width of Belt in inches.

Determining Proper Speed for Belt Conveyors

SPEEDS of Belt Conveyors depend upon the capacity desired;—the material being handled;—the amount of breakage allowable in discharging;—and the effect on the discharge chute. The speed should be as low as possible to safely carry the load with the belt kept full, but not less than 150 feet per minute when handling materials in bulk, except when used as a Picking or Sorting Belt.

Maximum Advisable Belt Speeds in feet per minute

Belt Width	14"	16"	18"	20"	24"	30"	36"	42"	48"
Maximum Speed	300	300	350	350	400	450	500	550	600

When Maximum Advisable Speeds are exceeded the load on the belt ordinarily has a tendency to thin out so that the capacity is not directly proportional to the speed.

Under 150 feet per minute speed the cost of the Belt Conveyor per ton of bulk materials handled even with a minimum ply of belt, commences to be uneconomical as compared with other types of conveyors equally suited to the operating conditions.

Speeds of Picking or Sorting Belts should be 40 to 50 feet per minute, but may be higher if the material is coarse, comparatively clean and the refuse easily discernible.

How to Figure Horsepower of a Belt Conveyor

A N exact formula embodying all the possible variables for arriving at the Horse Power of a conveyor is quite complex. A composite of accepted practice however has reduced it to the following:

For Horizontal Conveyors:—H.P. $= \dfrac{(CS+2.33\ T)L}{33000}$ at Drive Shaft.

For Inclined Conveyors:—H.P. $= \dfrac{(CS+2.33\ T)L}{33000} + \dfrac{TH}{990}$ at Drive Shaft.

C = Power Constant varying with the width of belt, from Table below.
S = Speed of Belt in feet per minute, page 130.
T = Load in Tons per hour. Use Capacity of Belts pages 129 and 130 at the speed chosen, unless operating conditions are such as to guarantee by a uniform rate of feed throughout the hour, that some smaller capacity will not be exceeded.
L = Length of Conveyor between centers in feet.
H = Vertical Height in feet that material is lifted, page 132.

Increase Horse Power thus obtained, 20% for conveyors under 50 feet centers, 10% for 50 to 100 feet, and 5% for 100 to 150 feet, due to the larger percentage of the total horse-power absorbed by the terminals of short conveyors as compared with those of longer centers. Now add for each fixed or movable tripper the Horse Power given in the following table. Increase this final Horse Power at the Drive or Head Shaft 5% for each speed reduction to line shaft, motor or engine, thru chains, belting or cut gears and 10% for each reduction through rough cast gears.

Width of Belt	C Power Constant	H. P. For Each Tripper	Width of Belt	C Power Constant	H. P. For Each Tripper
14"	.75	1	30"	2.45	2
16"	1.05	1	36"	3.55	2
18"	1.35	1½	42"	4.15	
20"	1.70	1½	48"	4.75	
24"	2.00	1½			

Tonnage Life of Belts satisfactory when Conveyors are Properly Installed

With one properly designed loading point and ⅛" good grade of rubber cover, a belt on a conveyor 100 feet long carrying coal or similar material, has been found in many cases to handle during its life, a tonnage equal to 500 times the square of the width of the belt in inches; two hundred feet long, twice as much, etc. This "Tonnage" is not to be taken as a limit to the performance of a good belt nor as a guarantee of performance in any case, but simply as a composite of what has been considered satisfactory service in a large number of cases.

Maximum Length of a Belt Conveyor is limited to the safe working tension of the belt when using the maximum number of plies permissible for proper flexibility. Ordinarily this length is about 450 to 500 feet under full loading. See Standard Belt Conveyors, pages 144 to 149.

Stretch. A good belt with a breaking strength of about 360 pounds per ply per inch of width is usually granted an allowance of 1% of its length for tightening including the permissible initial stretch incident to properly conforming to troughing carriers, driving pulleys and the load upon the belt.

The Larger the Curve from the Horizontal to the Incline the Better

IT is often desirable to install a belt conveyor partly on the horizontal and partly on an incline, with the change from the horizontal being made by an upward curve. Under such conditions it is necessary to know the smallest radius of curvature at which the belt will lie down upon the carriers when operating under conditions which give the greatest tendency to leave the carriers.

It is evident that such a tendency of the belt to rise from the carriers, as shown by dotted lines in the drawing below, will be greatest when the pull in the belt is greatest under the conditions of being completely loaded from "B" to "C" and empty on the curved portion from "C" to "D" up to "A." Under such conditions a pull equal to the strain at "C" will be in the empty curved portion of the belt along its entire length up to the driving pulley "A."

Now, if a rope or a belt "F, D, C, E, G" be freely suspended from the points "F" and "G" it will assume a curve known as a catenary, and the flatter the curve the greater will be the strain induced along the rope or belt by reason of its own weight independent of any material lying upon the same. It therefore follows that if an empty belt is so suspended as to assume the outline of that portion of a catenary in which the greatest induced strain along the curve from "D" to "C" is equal to the pulling strain at "C" then we will have obtained the smallest curve possible to meet the conditions of our problem.

This curve of the catenary, while mathematically quite complex, is closely approximated by a similar curve known as the parabola which may be quite easily plotted or laid out, since every point along the parabola is represented by the intersection of varying horizontal distances "X" measured in feet from the line "YY" with corresponding varying vertical distance "Y" measured from the line "XX" bearing the simple relation to each other of $X^2 = \frac{2T}{W} Y$.

Where W equals the weight in pounds per lineal foot of the empty belt and T equals the "horse power pull" or friction in pounds of the belt at "C". See pages 131 and 142.

A Radius of Curvature "K" may be taken in practice to approximate that portion "CD" of the parabolic curve as plotted from the above formula. This radius is ordinarily about 300 feet, with a smaller radius where a gradually increasing load becomes a uniform continuous load. For intermittent but full loading, a radius varying from 500 to 1000 feet may be required.

The Angle of Incline "M" should in all cases be about 10 or more degrees less than the angle of repose of the material on the belt. Therefore for the materials ordinarily handled on a belt the maximum incline may be taken at about 20 degrees. Many materials may be carried as high as 21°—23° and some few at 25°. Large lumps have a tendency to roll back upon the belt unless they are well intermixed with smaller pieces. Also an intermittent flow of most materials, near 20 degrees has a tendency to cause slipping and often the avalanching of all the material on the incline. Care should therefore be taken to insure a continuous stream, either large or small, of uniform sized or of well intermixed unsized materials.

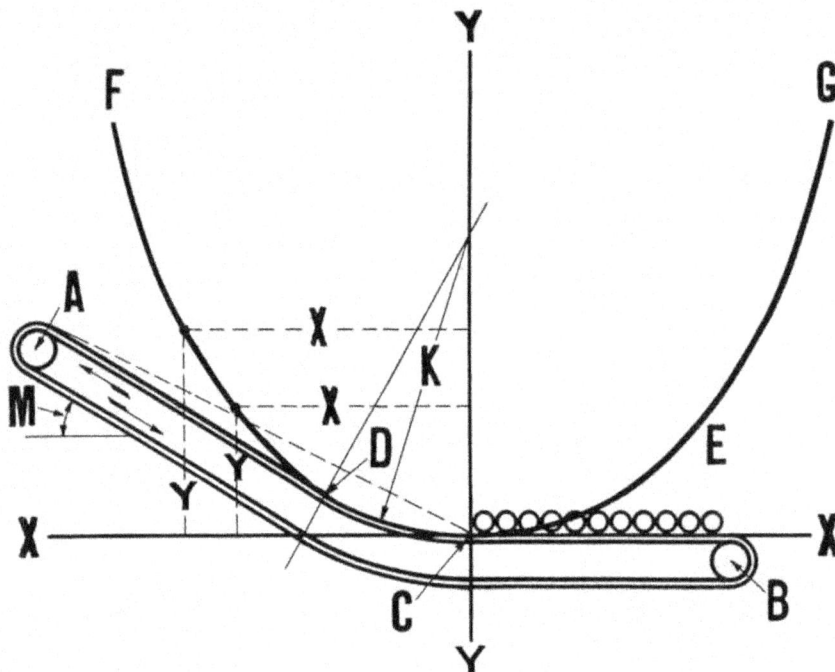

Diagram of Curve connecting the Horizontal and Incline of a Belt Conveyor

132

Maintaining a Steady Flow of Material to the Belt Loading Chute

MECHANICAL Loaders are often used to maintain a steady flow of material to the standard form of loading chute described below. Of the various types of loaders, we have found the use of the short screw conveyor for smaller materials to be the next step in simplicity following a hand controlled valve in the standard loading chute, while the continuous steel apron and reciprocating plate loaders have been found to be best suited to medium and larger size materials.

Loaders for Fine and Medium Size Materials

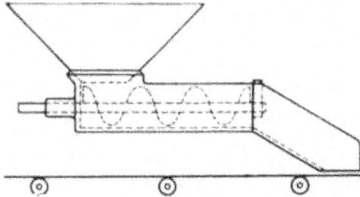

Fig. 1—Screw Conveyor Loader—Uniform Flow—Fast or Slow Speed.

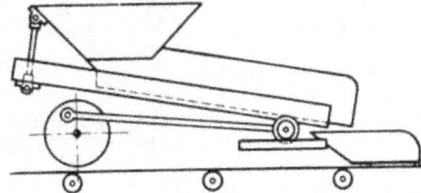

Fig. 2—Shaking Plate Loader—Approximate Uniform Flow—Short Strokes—Medium or High Speed.

Figure 1

Figure 2

Loaders for Medium and Large Size Materials

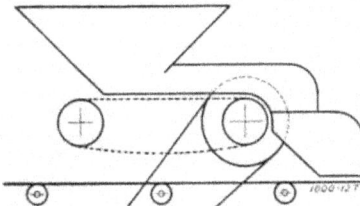

Fig. 3—Steel Apron Loader—Uniform Flow—Medium or Slow Speed.

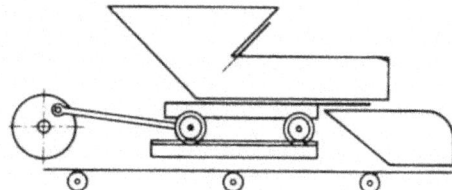

Fig. 4—Reciprocating Plate Loader—Intermittent but Uniform Flow—Medium Strokes—Slow Speed.

Figure 3

Figure 4

Lengthening Life of Belt by Loading Properly

LOADING is a very important feature to which due consideration should be given, as both the life of the belt and the capacity of the conveyor depend upon the way in which the conveyor is loaded. Heavy materials such as coal should never be allowed to strike the belt vertically. They should be baffled to at least a gentle drop onto the belt. For ideal loading the materials conveyed should be delivered to the center of the belt, in the direction of the belt travel and at as near the speed of the belt as possible thereby eliminating the wear of the belt due to slippage of material.

Figure 5—Protecting the Belt with Screenings

In designing a loading chute to approximate a speed of flow equal to the speed of the belt, provision should be made for as great an angle of the chute to the belt as conditions will permit, thereby making it readily possible to gradually place the chute at lesser angles to the belt until the proper speed of flow is obtained.

Fig. 6—Section thru Loading Skirt Boards

For heavy material intermixed with fines it is good practice to place a screen in the bottom of the loading chute, thereby allowing the fines to flow thru the screen to the belt and thus form a cushion for the large lumps, as shown in Fig. 5.

Skirt Boards. Materials conveyed should be at rest on the belt in from 6 to 8 feet from the loading point, therefore, skirt boards from 6" to 12" high and from 5 to 7 feet long are ordinarily required. For conveyors with the loading point at an angle add guard strips of rubber belting to the skirt boards, shown in Cross Section, Fig. 6.

Where W = Width of Belt in inches.

A = ⅝W. for Belts 14 and 16 inches wide and B = 1⅜"

A = ⅔W. less 1 inch for Belts 18 and 20 inches wide and B = 1¾".

A = ⅔W. less 2 inches for Belts 24 inches and wider and B = 1¾".

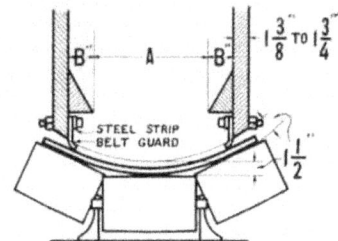

Three Pulley Troughing Belt Carriers

THE diameter and troughing angle of Jeffrey Belt Carriers are the result of extensive engineering experience in the successful handling of various classes of materials. Exceedingly strong, of medium weight and well machined. The feature of overlapping the pulleys in this type of carrier is a big element in preventing undue creasing of the belt in the trough. All standard carriers are equipped with a High Pressure Lubrication System which insures thorough lubrication of all working surfaces.

IMPORTANT:—Completely fill the grease passages when first installed and an occasional application of grease will keep all parts well lubricated. To insure the belt running true and central, it should always approach the carrier from the side upon which the horizontal pulley is placed.

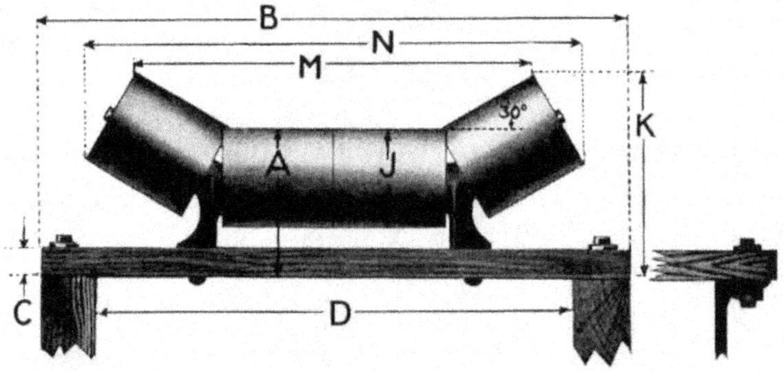

Three Pulley Carrier can be Mounted upon Wood or Steel Stringers

Dimensions of Three Pulley Troughing Idlers

Width* of Belt Inches	List Price	Approx. Weight with Boards Lbs.	Dimensions—Inches							
			A	B	C	D	J	K	M	N
14	See	**31**	**7⅜**	**24**	**1⅛**	**20**	**5**	**9⅝**	**14**	**19**
16	Price	**33**	**7⅜**	**26**	**1⅛**	**22**	**5**	**10⅛**	**15¾**	**20¾**
18	List	**44**	**8⅜**	**30**	**1⅛**	**24**	**6**	**11⅝**	**17½**	**23½**
20	Bulletin	**49**	**8⅜**	**32**	**1⅛**	**26**	**6**	**11⅝**	**19¾**	**25¾**

* All Carriers **in Bold Face Type** are **Carried in Stock** in quantities to meet all ordinary demands.
For general service above 20 inch belts use Five Pulley Carriers, page 135.

Five-Pulley Troughing Belt Carriers

"The Ideal Carrier"

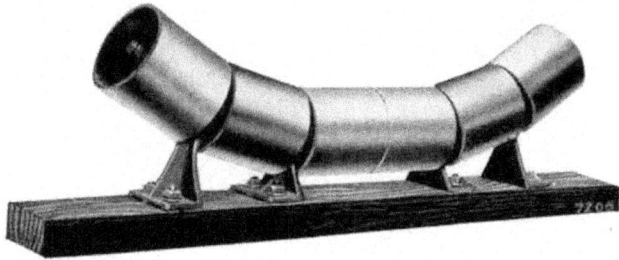

THE Jeffrey Five Pulley Carrier embodies all the high qualities of the Standard Three Pulley Carrier, and in addition permits of a closer conformity to the natural troughing effect of the belt, for the carrying of coal, sand, crushed stone, and other heavy bulk materials such as ore, rock, earth, etc.

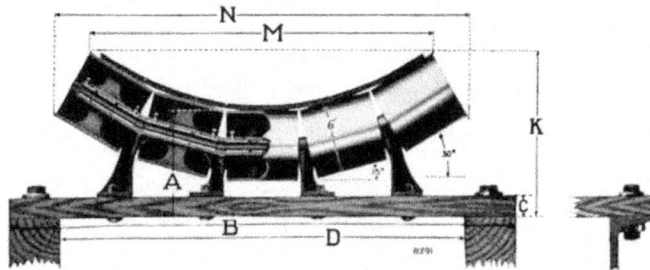

In the Five-Pulley Carrier the Pulleys are set in line upon hollow renewable steel spindles connecting four rigid and well proportioned supporting stands. By this arrangement an exceedingly rigid construction is secured, with the spindles serving as a continuous tube through which grease is forced to all the pulleys by means of the "High Pressure Lubrication System".

IMPORTANT:—Completely fill the grease passages when first installed, then an occasional application of grease will keep all parts well lubricated.

Dimensions of Five Pulley Troughing Belt Carriers

Width* of Belt Inches	List Price	Approx. Weight Board Lbs.	Dimensions—Inches						
			A	B	C	D	K	M	N
24	See	70	8⅞	36	1⅝	30	12⅛	25	31
30	Price	82	8⅞	44	1⅝	38	12⅞	30	36
36	List	99	8⅞	50	1⅝	44	13⅝	36¼	42¼
42	Bulletin	115	8⅞	56	1⅝	50	13¾	41⅞	47⅞
48		130	8⅞	62	1⅝	56	15¼	47	53

*All Carriers are **Carried in Stock** in quantities to meet all ordinary demands.

Concentrating Belt Carriers

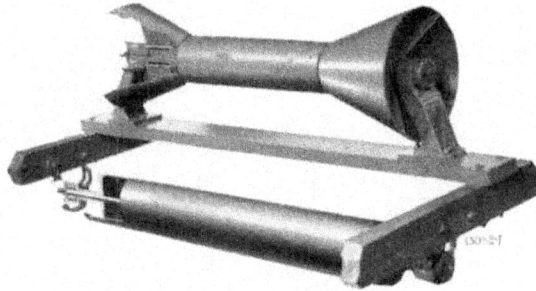

Concentrating Carrier

At the left is shown the Jeffrey Concentrating Carrier, made up of cast iron cone and steel flat pulleys, mounted upon Roller Bearings and equipped with a high pressure lubrication system.

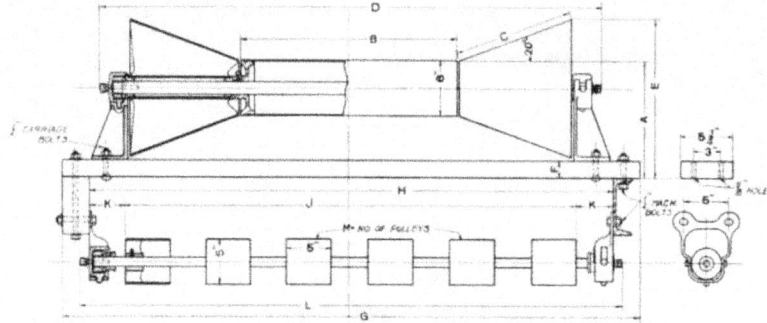

Dimensions in Inches of Belt Carriers Handling 50 lbs. per Cu. Ft.

Width of Belt Inches	Approx. Weight Lbs.	Capacity at 100 F. P. M.		A	B	C	D	E	F	G	H	J	K	L	M
		In cu ft. per Hour	In Tons per Hour												
24	110	2016	50.4	$10\frac{1}{4}$	$11\frac{5}{8}$	$6\frac{7}{8}$	$32\frac{5}{8}$	$12\frac{5}{8}$	$1\frac{5}{8}$	38	32	25	$3\frac{1}{2}$	35	4
30	125	3150	78.8	$10\frac{11}{16}$	$14\frac{5}{8}$	$8\frac{5}{16}$	$38\frac{3}{8}$	$13\frac{1}{2}$	$1\frac{5}{8}$	44	38	31	$3\frac{1}{2}$	41	5
36	155	4536	113.4	$11\frac{1}{4}$	$17\frac{5}{8}$	$9\frac{7}{8}$	$44\frac{1}{4}$	$14\frac{5}{8}$	$1\frac{5}{8}$	50	44	37	$3\frac{1}{2}$	47	5
42	185	6174	154.4	12	$20\frac{5}{8}$	$11\frac{3}{16}$	50	$15\frac{7}{8}$	$1\frac{7}{8}$	58	52	43	$4\frac{1}{2}$	55	6
48	205	8064	201.6	$12\frac{9}{16}$	$23\frac{5}{8}$	13	$56\frac{5}{8}$	17	$1\frac{7}{8}$	64	58	49	$4\frac{1}{2}$	61	6
54	250	10250	256.3	$13\frac{1}{2}$	$26\frac{5}{8}$	$14\frac{5}{16}$	$61\frac{5}{8}$	$18\frac{3}{8}$	$2\frac{3}{8}$	69	63	55	4	66	7
60	285	12950	324.0	14	$29\frac{5}{8}$	$15\frac{13}{16}$	$67\frac{1}{2}$	$19\frac{5}{8}$	$2\frac{3}{8}$	75	69	61	4	72	8
For Handling 150 Lbs. per Cu. Ft.															
24	110	2016	151.2	$10\frac{1}{2}$	$11\frac{5}{8}$	$6\frac{7}{8}$	$32\frac{5}{8}$	$12\frac{7}{8}$	$1\frac{7}{8}$	38	32	25	$3\frac{1}{2}$	35	4
30	125	3150	236.4	$10\frac{15}{16}$	$14\frac{5}{8}$	$8\frac{5}{16}$	$38\frac{3}{8}$	$13\frac{3}{4}$	$1\frac{7}{8}$	44	38	31	$3\frac{1}{2}$	41	5
36	155	4536	340.2	$11\frac{1}{2}$	$17\frac{5}{8}$	$9\frac{7}{8}$	$44\frac{1}{4}$	$14\frac{7}{8}$	$1\frac{7}{8}$	50	44	37	$3\frac{1}{2}$	47	6
42	220	6174	463.2	$12\frac{1}{2}$	$20\frac{5}{8}$	$11\frac{3}{16}$	50	$16\frac{5}{8}$	$2\frac{3}{8}$	58	52	43	$4\frac{1}{2}$	55	6
48	245	8064	604.8	$13\frac{1}{16}$	$23\frac{5}{8}$	13	$56\frac{5}{8}$	$17\frac{1}{2}$	$2\frac{3}{8}$	64	58	49	$4\frac{1}{2}$	61	6
54	305	10250	768.9	$13\frac{3}{4}$	$26\frac{5}{8}$	$14\frac{5}{16}$	$61\frac{5}{8}$	$18\frac{5}{8}$	$2\frac{5}{8}$	69	63	55	4	66	7
60	345	12950	972.0	$14\frac{1}{4}$	$29\frac{5}{8}$	$15\frac{13}{16}$	$67\frac{11}{16}$	$19\frac{5}{8}$	$2\frac{5}{8}$	75	69	61	4	72	8

High Pressure Lubrication System

The Fitting   **The Gun**

THE High Pressure Lubrication System gives assurance that the center idler receives its quantity of the lubricant as well as the end ones. With the aid of the gun first pack the tube with lubricant. Refill the gun and then, after working up pressure in the gun, thrust the nozzle into the fitting, which automatically releases the stored up pressure and instantaneously jolts the packed lubricant with a blow of tremendous force, driving the lubricant out through every available opening.

High Pressure Lubrication System (Cont'd)

THE grease in being driven out by this explosive action carries with it any gritty and dirty particles harmful to the working surfaces and also opens up any clogged passages.

Ordinarily it is not necessary to clean off the bearing surfaces if the system is used at regular intervals, but if for any reason they should become clogged or gritty, several applications of kerosene from the gun will thoroughly clean the surfaces. This is made possible by the fact that the gun handles effectively all grades of grease and any grade of oil. It is of rugged construction, and as nearly acid proof as possible.

The sturdily constructed fitting is especially suited for the hard service encountered in elevating and conveying work. It is equipped with a hinged dust tight cap.

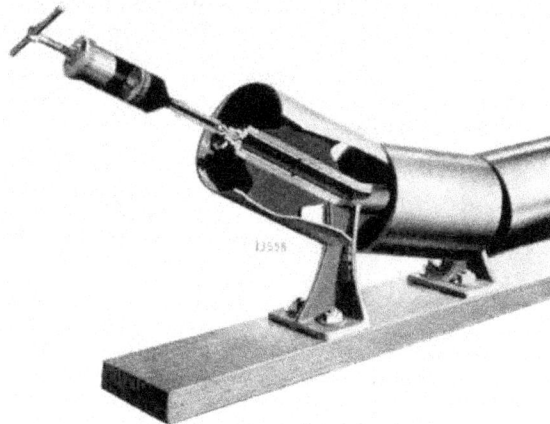

Showing High Pressure System applied to the Regular Five Pulley Type of Idler.

The system is quick and clean to operate. There is no complicated nozzle locking device for attaching gun to fitting, and dust cap can not be lost.

Belt Conveyor Guide Pulleys

For Three and Five Pulley Troughing Belt Carriers

Jeffrey Guide Pulleys are designed so as to permit of a minimum over-all width of conveyor. The smooth, curved ends of the Jeffrey patented pulleys protect the edges of the belt from possible injury, whereas the old style square-edged pulleys ruined many belts.

Guide Pulleys are seldom necessary for belts over 24" and cannot be used for any Conveyor using a Tripper.

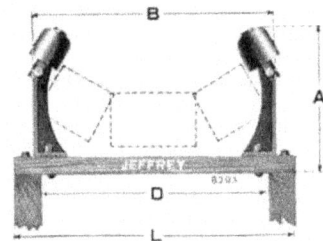

Guide Pulleys for Three Pulley Carriers

Width* of Belt Inches	List Price	Approx. Weight with Board, Lbs.	Dimensions—Inches			
			A	B	D	L
14		21.0	13¾	22¾	20	24
16	See	21.5	13¾	25	22	26
18	Price	23.0	15¾	26¼	24	30
20	List	25.0	15¾	28½	26	32
24	Bulletin	25.8	16¾	33	30	36
30		27.5	16¾	39¾	38	44

For Five Pulley Carriers

24	See	25.8	16¾	33¼	30	36
30	Price List	30.5	16⅞	38½	38	44
36	Bulletin	35.0	17¼	45¼	44	50

* All Guide Pulleys in Bold Face Type are **Carried in Stock** in quantities to meet all ordinary demands.

Belt Conveyor Return Idlers
With Swivel Bearings and Grease Fittings for Three and Five Pulley Carriers

Can be mounted on either wood or steel stringers.

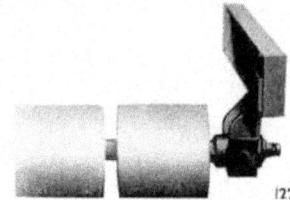

Bottom-Hanging Swivel-Bearing

Side-Hanging type furnished unless otherwise specified.

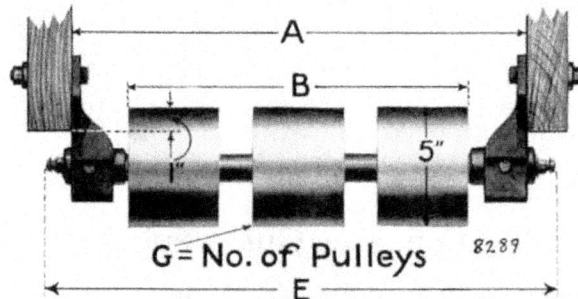

Side-Hanging Swivel-Bearing

G= No. of Pulleys

Dimensions of Return Idlers

Width of Belt Inches	List Price Each	Approx. Weight Lbs.	A For both Side and Bottom Hangers	B	E For Side Hangers	E For Bottom Hangers	G
14		23	20	15	23¾	27½	3
16	See	23	22	17	25¾	29½	3
18		28	24	19	27¾	32	4
20	Price	29	26	21	29¾	34	4
24		38	30	25	33⅞	38⅛	4
30	List	47	38	31	41⅞	46⅛	5
36		48	44	37	47⅞	52⅛	5
42	Bulletin	56	50	43	53⅞	58⅛	6
48		58	56	49	59⅞	64⅛	6

Idlers in **Bold Face Type** are **Carried in Stock.**

The Jeffrey Malleable Iron Hanger while lighter in appearance than the ordinary cast iron hanger is much stronger in service and therefore a greater insurance to the continuous performance of the conveyor in rough and rugged service.

All edges are nicely rounded to prevent any possible injury to the belt.

Automatic Traveling Trippers

(Pat. applied for)

**Self-
Propelled
Auto-
matic
Reversing
Belt
Tripper**

**Hand-
Propelled
Belt
Tripper**

T RIPPERS must be used where it is necessary to discharge the load from the belt at intermediate points along the length of the conveyor. If the discharge is at one fixed point a Stationary Tripper may be used.

To discharge at a number of points along a conveyor, it used to be customary to install a number of stationary trippers with a chute and a valve at each point so arranged that the material carried could be loaded back onto the belt and be carried to the next tripper. However, as the life of a belt is shortened in proportion to the number of loading points onto it, it readily can be seen that this method was far from being ideal. For such conditions we recommend one of the Traveling Trippers shown on this page.

Automatic Traveling Trippers

(Pat. applied for)

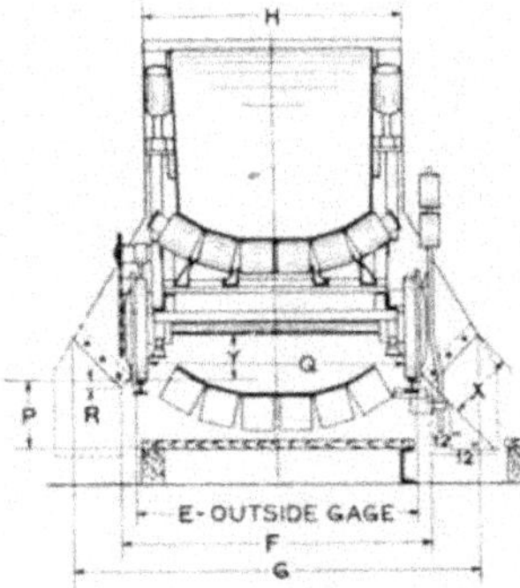

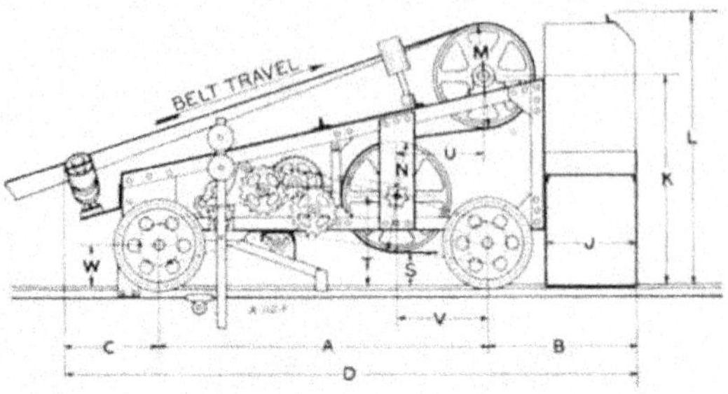

Self-Propelled Automatic Reversing Belt Tripper

(Right hand shown)

Hand-Propelled Belt Tripper

(Right hand shown)

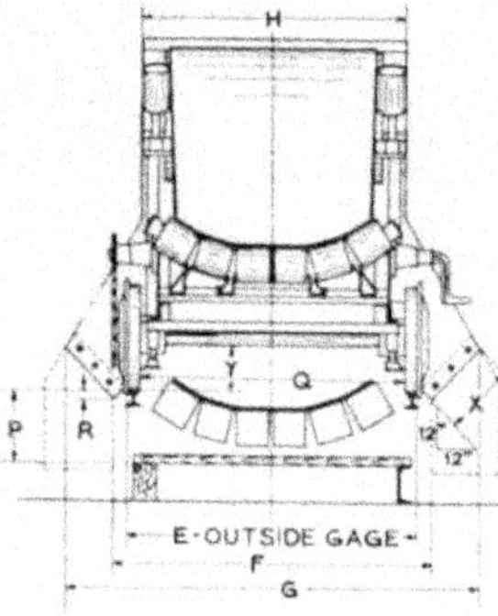

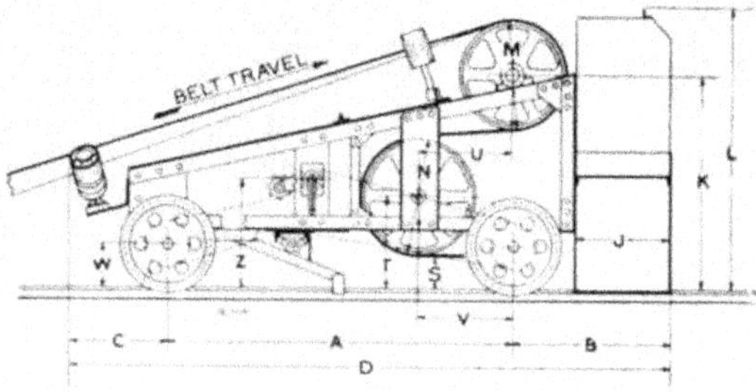

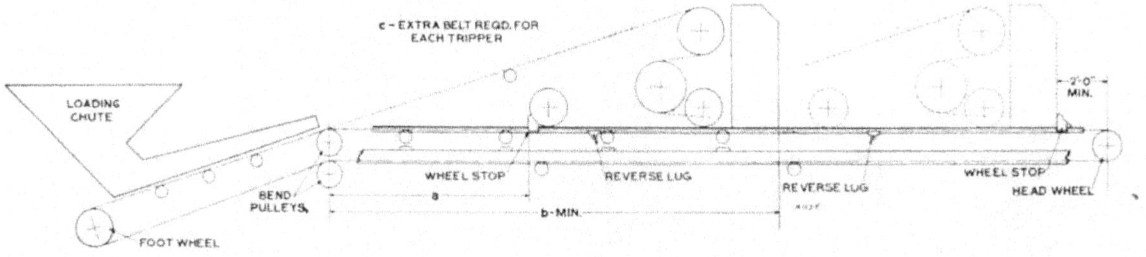

General Arrangement of Belt Trippers

Dimensions of Hand and Self-Propelled Belt Trippers

Width of Belt	A	B	C	D	E	F	G	H	J	K	L	M	N	P
18"	5'-5"	2'-2"	19"	9'- 2"	2'-8"	3'-5"	4'-7"	2'- 3½"	14"	3'-0"	4'- 0"	16"	18"	9⅜"
24"	5'- 7"	2'-4"	20"	9'- 7"	3'-2"	3'-9"	5'-1"	2'- 9½"	16"	3'-2"	4'- 2½"	18"	20"	9⅝"
30"	5'- 9½"	2'-6"	20"	9'-11½"	3'-8"	4'-1"	5'-6"	3'- 4"	18"	3'-5"	4'- 6¾"	20"	22"	12¾"
36"	6'-1¼"	2'-9"	20½"	10'- 6¾"	4'-2"	4'-7"	6'-0"	3'-10"	20"	3'-8"	4'-10½"	22"	24"	12¾"

Width of Belt	Q	R	S	T	U	V	W	X	Y	Z	a	b	c
18"	2'- 5"	1¼"	7"	16½"	17"	15½"	9"	10"	7½"	21½"	7'-0"	15'- 6"	8'-0"
24"	2'-11"	2"	7¼"	17¾"	17¼"	16¼"	9"	11"	7½"	21½"	7'-0"	15'-10"	8'-6"
30"	3'- 3"	½"	7½"	19"	17¾"	17¼"	9"	12"	7½"	21¾"	7'-0"	16'- 3"	9'-0"
36"	3'- 9"	1"	6½"	19"	18½"	20"	9"	12"	7½"	21¾"	7'-0"	16'- 9"	10'-0"

Unless otherwise specified, trippers are furnished right hand as shown. A right hand tripper has the operating mechanism on the right hand side looking in the direction of belt travel.

Trippers can be fitted with cleaning brushes if so desired.

When ordering Standard Belt Conveyor, specify extra belt as given by "c" in table.

Do not use Guide Idlers with either of the Trippers.

Add 3 to 4 inches to dimensions given above for proper clearance to any outside structure.

Support Tripper rails midway between troughing carriers with rail joints over such supports

Bend Pulleys extend the Travel of the Tripper

THE longest possible travel of a Tripper over its conveyor is obtained by deflecting the Foot End of the Belt Conveyor over Bend or Snub Pulleys, so that the angle of the Standard Loading Chute is the same as the angle "R" of the tripper, which is approximately 18°. By this method a maximum distribution of material by the tripper is secured for a comparatively short conveyor, a feature which is highly desirable when operating in a limited space over a short storage bin.

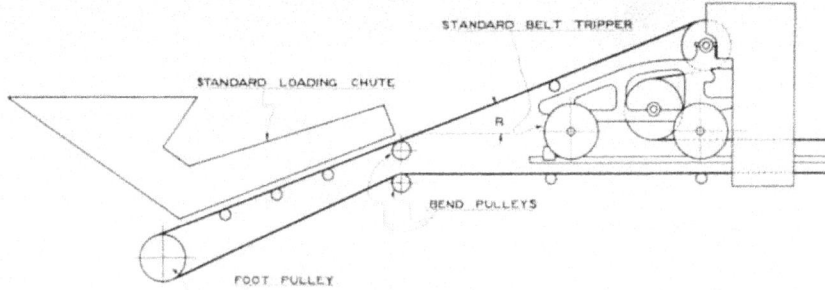

In this connection the hopper part of the Standard Loading Chute may be replaced by the head chute of an elevator which brings the material handled up from a track hopper below, as in a Power Plant, a Retail Coal Pocket, or a Sand and Gravel Plant.

It is to be noted that the skirt boards of the Loading Chute do not extend beyond the upper bend pulley.

When the Tripper does not approach near enough to the loading chute for the belt in its gradual rise from the carriers to be cut by the loading skirts, the conveyor and its loading chute may be installed, wholly on the horizontal. This condition applies especially to very long storage spaces.

Brushes Help Preserve the Carrying Surface of the Belt

Jeffrey Belt Cleaning Brush

WHEN there is any tendency for particles of the material handled to adhere to the surface of a belt, a rotating cylindrical cleaning brush should be used. Where space will permit, use a brush 12 inches in diameter but never less than 8 inches, with the face of the brush equal to the belt width.

The speed in feet per minute of the surface of the brush should be at least 800 to 1000 for dry dusty materials; 1000 to 1200 for damp materials; and 1200 to 1500 for wet sticky materials. Such brushes are specially constructed to embody to a nicety, stiffness, flexibility and durability for a maximum of clean sweeping without injury to the belt.

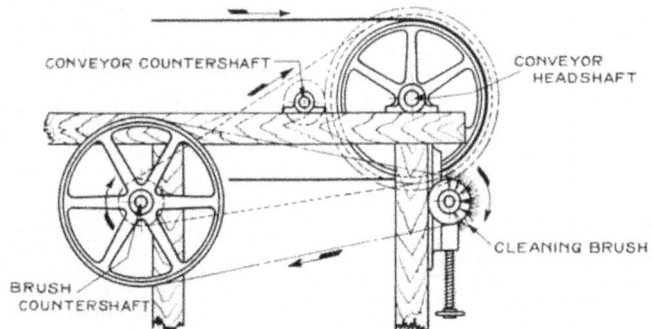

Cleaning Brush Arrangement for Head Pulley

Conveyor Belting

"Century" Conveyor Belt

Typical Cross-section of "Century" Conveyor Belt, 4-Ply $\frac{1}{16}$" Cover

THE "Century" Belt is made exclusively for us. The DUCK is of more than ordinary tensile strength longitudinally, and admits of great flexibility cross-wise thereby giving a close conformity to the troughing carriers and insuring maximum capacities.

The "FRICTION" or adhesive between the plies is a good substantial rubber compound of strong elastic tendrils, like threads, which hold the plies together and keep their life under proper working conditions during the service of the belt.

The COVER is strong, tenacious, and resilient. It protects the body of the belt from the entrance of moisture; cushions ordinary impact without injury; and reduces wear from abrasion to a minimum where proper loading facilities are installed.

The EDGES will stay on until the belt is worn through. The top cover in one piece is carried around the edges and into the back cover, where its ends are connected into the belt structure. Our belts are cured under stretch. This avoids any troublesome skew of the belt when put in service, and makes a belt which will run straight and stay straight.

Century Rubber Belts are adapted to the handling of any materials either wet or dry, which are not of a plastically sticky nature. Some semi-adhesive substances however may be handled where rotating brushes, especially designed for such service, are used. Materials hotter than 140 degrees—150 degrees F. (60 degrees, to 66 degrees C) will too rapidly deteriorate rubber belts, and therefore should be reduced in temperature by baffle chutes or other means leading into the loading chute, before touching the belt.

A Belt should conform to a troughing carrier by its own weight in order to get the guiding action of the central horizontal pulleys of the carrier. If too stiff the belt will ride the inclined sides of the troughing pulleys or run out of line over the edges of the pulleys, thereby injuring the edges of the belt. If too flexible the belt will crease lengthwise in the bends of the carrier trough and be weak at the edges.

For List Price—See Price List Bulletin

PLY	Rubber Cover Top Side	Weight in Pounds per Lineal Foot of Belts of the Following Widths										Thickness
		12"	14"	16"	18"	20"	24"	30"	36"	42"	48"	
3	$\frac{3}{128}$"	1.27	1.48	1.69	1.90	2.11	2.53	3.16	3.79	4.42	5.05	$\frac{7}{32}$"
	$\frac{1}{16}$"	1.51	1.76	2.01	2.26	2.51	3.01	3.76	4.51	5.26	6.01	$\frac{1}{4}$"
	$\frac{1}{8}$"	1.99	2.32	2.65	2.98	3.31	3.97	4.96	5.95	6.94	7.93	$\frac{5}{16}$"
	$\frac{3}{16}$"	2.47	2.88	3.29	3.70	4.11	4.93	6.16	7.39	8.62	9.85	$\frac{3}{8}$"
4	$\frac{3}{128}$"	1.54	1.79	2.05	2.30	2.56	3.07	3.83	4.60	5.36	6.13	$\frac{9}{32}$"
	$\frac{1}{16}$"	1.78	2.07	2.37	2.66	2.96	3.55	4.43	5.32	6.20	7.09	$\frac{5}{16}$"
	$\frac{1}{8}$"	2.26	2.63	3.01	3.38	3.76	4.51	5.63	6.76	7.88	9.01	$\frac{3}{8}$"
	$\frac{3}{16}$"	2.74	3.19	3.65	4.10	4.56	5.47	6.83	8.20	9.56	10.93	$\frac{7}{16}$"
5	$\frac{3}{128}$"	1.80	2.10	2.40	2.70	3.00	3.60	4.50	5.40	6.30	7.20	$\frac{11}{32}$"
	$\frac{1}{16}$"	2.04	2.38	2.72	3.06	3.40	4.08	5.10	6.12	7.14	8.16	$\frac{3}{8}$"
	$\frac{1}{8}$"	2.52	2.94	3.36	3.78	4.20	5.04	6.30	7.56	8.82	10.08	$\frac{7}{16}$"
	$\frac{3}{16}$"	3.00	3.50	4.00	4.50	5.00	6.00	7.50	9.00	10.50	12.00	$\frac{1}{2}$"
6	$\frac{3}{128}$"	2.07	2.41	2.76	3.10	3.45	4.14	5.17	6.21	7.24	8.28	$\frac{13}{32}$"
	$\frac{1}{16}$"	2.31	2.69	3.08	3.46	3.85	4.62	5.77	6.93	8.08	9.24	$\frac{7}{16}$"
	$\frac{1}{8}$"	2.79	3.25	3.72	4.18	4.65	5.58	6.97	8.37	9.76	11.16	$\frac{1}{2}$"
	$\frac{3}{16}$"	3.27	3.81	4.36	4.90	5.45	6.54	8.17	9.81	11.44	13.08	$\frac{9}{16}$"
7	$\frac{3}{128}$"	2.33	2.72	3.11	3.50	3.89	4.67	5.84	7.01	8.18	9.35	$\frac{15}{32}$"
	$\frac{1}{16}$"	2.57	3.00	3.43	3.86	4.29	5.15	6.44	7.73	9.02	10.31	$\frac{1}{2}$"
	$\frac{1}{8}$"	3.05	3.56	4.07	4.58	5.09	6.11	7.64	9.17	10.70	12.23	$\frac{9}{16}$"
	$\frac{3}{16}$"	3.53	4.12	4.71	5.30	5.89	7.07	8.84	10.61	12.38	14.15	$\frac{5}{8}$"
8	$\frac{3}{128}$"	2.60	3.03	3.47	3.90	4.34	5.21	6.51	7.82	9.12	10.43	$\frac{17}{32}$"
	$\frac{1}{16}$"	2.84	3.31	3.79	4.26	4.74	5.69	7.11	8.54	9.96	11.39	$\frac{9}{16}$"
	$\frac{1}{8}$"	3.32	3.87	4.43	4.98	5.54	6.65	8.31	9.98	11.64	13.31	$\frac{5}{8}$"
	$\frac{3}{16}$"	3.80	4.43	5.07	5.70	6.34	7.61	9.51	11.42	13.32	15.23	$\frac{11}{16}$"

For Troughed Belts:—Between Heavy Zig Zag Lines, Standard Ply for Proper Flexibility.

For Flat Belts:—All belts below upper Zig Zag Lines are Standard for Proper Flexibility, but for very light service 3 ply may be used for 16" and 18" with 4 ply for 24" belts.

Belt Covers best adapted to Light and Heavy Service

Rubber Covers:—For clay, sawdust, shavings, etc., use "Regular Cover" (About $\frac{3}{128}$ inches thick); For cement, small coal, dirt, sand, etc. $\frac{1}{16}$" cover; and for cold clinker, ores, stone, large coal, etc. $\frac{1}{8}$" cover. At purchaser's request $\frac{3}{16}$" and $\frac{1}{4}$" covers are furnished for very severe service.

Conveyor Belting

BALATA BELT occupies a position between Rubber and Stitched Canvas Belting described below. It is adapted to the handling of non-abrasive and semi-gritty materials under dry or wet conditions, at temperatures not exceeding 120 degrees Fahrenheit.

Balata is a vegetable gum found in Venezuela and the Dutch East Indies. In nature it lies between gutta percha and india rubber, but differs from them in its great tensile strength, freedom from oxidation, and the fact that it does not deteriorate with age. The Balata in a liquid form, is applied under pressure to the fabric, so that the gum penetrates every fibre of the fabric, thoroughly water proofing it.

Jeffrey Stitched Canvas Belting is suited to the handling of non-abrasive and semi-gritty materials under dry or wet conditions, at temperatures, not exceeding 212 degrees Fahrenheit.

A special width of high grade cotton duck is woven for each width and ply of belting, thus giving two selvage edges, thereby insuring true and even running on the carriers. Every belt is stitched lengthwise with heavy cotton twine in rows about one-quarter inch apart, each row being perfectly straight for the entire length of the belt. The complete belts are immersed and saturated in a compound which renders them impervious to the action of water, steam, oils and gases, but does not affect their flexibility.

Made in 4, 5, 6 and 8 ply in 12", 14", 16", 18" to 48" etc. widths.

Cotton Belting—The strength of this belt is equal to that of rubber or canvas, combined with exceptional flexibility; thus making it an excellent belt, for handling light non-abrasive materials under dry conditions.

Cotton belting being solid woven, under a constant stress, the pull is distributed equally thru out all parts with no plies to separate.

Made in 2, 3, 4, 5, 6 and 8 ply in 14", 16" to 48", etc. widths.

Lacing. For all ordinary belt conveyor installations a flexible metallic lacing with teeth which clinch around the warp or lengthwise threads of the belt and not around the filler threads, should be used such as:—the "Alligator," "Turtle" and other similar brands. In lacing a belt be sure to first make the belt ends square with the sides.

"Maxlife" Conveyor Belt

"MAXLIFE" Belting embodies all of the high class construction features of the "Century" brand, plus an extra quality of rubber, both in toughness and wearing qualities for hard abrasive materials, and especially in such service where properly designed loading facilities, noted on page 133, cannot completely be attained or maintained with the "Century" brand.

This Belt while of par excellence for any material handled on a belt has its special or economical application in the Metal Mining Industry where the service is extremely hard and the tonnage large; in fact it was for that industry that "Maxlife" Belting was designed and built, after a most careful field analysis of all the elements entering into the handling of ores.

For List Price—See Price List Bulletin

1/16", 1/8", 3/16", 1/4" Cover						1/16", 1/8", 3/16", 1/4" Cover						1/16", 1/8", 3/16", 1/4" Cover					
Width Inches	4 Ply	5 Ply	6 Ply	7 Ply	8 Ply	Width Inches	4 Ply	5 Ply	6 Ply	7 Ply	8 Ply	Width Inches	4 Ply	5 Ply	6 Ply	7 Ply	8 Ply
12	*					24	*	*	*			36		*	*	*	*
14	*					26	*	*	*			42		*	*	*	*
16	*	*				28	*	*	*			48		*	*	*	*
18	*	*				30	*	*	*			54		*	*	*	*
20	*	*	*			32			*	*	*	60		*	*	*	*
22	*	*	*			34			*	*	*						

Bold Faced Type indicates Belt Widths for Standard Carriers. Thickness of Cover at head of column applies to carrying side only. Belts over 450 to 500 feet long are furnished in 2 lengths, with the shorter pieces not less than 100 feet long.

*Indicates the plys of belt which can be furnished.

Horizontal Standard Belt Conveyors 0 to 100 Feet Centers

For 50 Pound Materials Such as Coal

No. of Conveyor	301	302	303	304	306	309	312	421	422
Size of Material, Inches (page 129)									
Uni-form or Size } 70% to 80% of Unsized Material:	2	2½	3	3½	4½	6	7½	9	10½
With Largest Pieces not to exceed 10% of all	3	4	5	6	8	11	14	17	20
Capacity—(pages 129-130)									
Tons per Hour	36	48	70	86	143	254	404	601	862
Speed in Feet per Minute	225	225	260	260	300	340	375	410	450
Belt—"Century" (page 142)									
Width—Inches	14	16	18	20	24	30	36	42	48
Ply, with $\frac{1}{16}''$ Rubber Cover	4	4	4	4	5	5	6	6	7
Spacing of Carriers									
Three Pulley Troughing—Inches (page 134)	60	60	54	54	....	....	....	....	....
Five Pulley Troughing—Inches (page 135)	....	....	....	....	48	48	42	42	42
Side Hanging Returns, Feet (page 138)	10	10	10	10	10	10	10	10	10
Head Shaft. At Discharge End									
Diameter of Shaft—Inches	1 15/16	1 15/16	1 15/16	1 15/16	2 7/16	2 7/16	2 15/16	2 15/16	3 7/16
Diameter of Pulley—Inches	20	20	20	20	24	24	30	30	36
Diameter of Gear—Inches	23.89	23.89	23.89	23.89	23.89	23.89	29.83	29.83	40.12
Pitch of Gear—Inches	1	1	1	1	1	1	1¼	1¼	1½
Face of Gear—Inches	2½	2½	2½	2½	2½	2½	3	3	4
Counter Shaft									
Diameter of Shaft—Inches	1 7/16	1 7/16	1 7/16	1 7/16	1 15/16	1 15/16	2 7/16	2 7/16	2 11/16
Rev. per Minute	202	202	234	234	225	254	240	262	268
Diameter of Pinion—Inches	5.12	5.12	5.12	5.12	5.12	5.12	6.01	6.01	7.22
Horse Power (page 131) at Counter Shaft for 100 Feet Centers	0.9†	1.2†	1.8†	2.2	3.3	4.8	7.7	10.6	14.2
Diameter of Foot Shaft—Inches	1 7/16	1 7/16	1 7/16	1 7/16	1 15/16	1 15/16	2 7/16	2 11/16	2 11/16
Diameter of Foot Pulley—Inches	16	16	16	16	20	20	24	24	28
Approx. Weights—Lbs.									
Terminals§	635	674	727	777	1182	1358	2126	2505	3648
Conveyor per Ft. Centers§	13.0	14.0	18.6	20.4	28.9	34.8	47.0	55.0	63.0
Guide Idlers per Set (page 137)	*	*	*	*	25.8	30.5	35.0	....	....

*Guide Idlers for these conveyors are included in weight of "Conveyor per Ft. Centers."

†In no case should separate motor drives be less than 1½ to 2 Horse Power.

§Terminals comprise Head, Counter and Foot Shafts, Bearings, Set Collars, Gear and Pinion; Head and Foot Pulleys with Belt to extend halfway around pulleys. Conveyor per Foot Centers comprise carrying and return idlers with the necessary belt.

For Erection Dimensions of above Conveyors, see page 150.

Horizontal Standard Belt Conveyors 101 to 200 Feet Centers

For 50 Pound Materials Such as Coal

No. of Conveyor	313	314	315	316	318	321	324	426	427
Size of Material, Inches (page 129)									
Uniform Size or 70% to 80% of Unsized Material	2	2½	3	3½	4½	6	7½	9	10½
With Largest Pieces not to exceed 10% of all	3	4	5	6	8	11	14	17	20
Capacity, (pages 129–130)									
Tons per Hour	36	48	70	86	143	254	404	601	862
Speed in Feet per Minute	225	225	260	260	300	340	375	410	450
Belt, "Century" (page 142)									
Width—Inches	14	16	18	20	24	30	36	42	48
Ply, with $\frac{1}{16}$" Rubber Cover	4	4	4	4	5	5	6	6	7
Spacing of Carriers									
Three Pulley Troughing—Inches (page 134)	60	60	54	54	....	....	....	....	....
Five Pulley Troughing—Inches (page 135)	....	....	....	....	48	48	42	42	42
Side Hanging Returns, Feet (page 138)	10	10	10	10	10	10	10	10	10
Head Shaft. At discharge end									
Diameter of Shaft—Inches	$1\frac{15}{16}$	$1\frac{15}{16}$	$1\frac{15}{16}$	$1\frac{15}{16}$	$2\frac{7}{16}$	$2\frac{15}{16}$	$3\frac{7}{16}$	$3\frac{7}{16}$	$3\frac{15}{16}$
Diameter of Pulley—Inches	20	20	20	20	24	24	30	30	36
Diameter of Gear—Inches	23.89	23.89	23.89	23.89	23.89	29.83	32.00	36.78	36.78
Pitch of Gear—Inches	1	1	1	1	1	1¼	1½	1¾	1¾
Face of Gear—Inches	2½	2½	2½	2½	2½	3	4	5½	5½
Counter Shaft									
Diameter of Shaft—Inches	$1\frac{7}{16}$	$1\frac{7}{16}$	$1\frac{7}{16}$	$1\frac{7}{16}$	$1\frac{15}{16}$	$2\frac{7}{16}$	$2\frac{11}{16}$	$2\frac{11}{16}$	$2\frac{11}{16}$
Rev. per Minute	202	202	234	234	225	270	214	245	226
Diameter of Pinion—Inches	5.12	5.12	5.12	5.12	5.12	6.01	7.22	7.86	7.86
Horse Power (page 131) at Counter Shaft for 200 Feet Centers	1.8†	2.4	3.6	4.4	6.6	9.6	15.4	21.2	28.4
Diameter of Foot Shaft—Inches	$1\frac{15}{16}$	$1\frac{15}{16}$	$1\frac{15}{16}$	$1\frac{15}{16}$	$1\frac{15}{16}$	$2\frac{7}{16}$	$2\frac{7}{16}$	$2\frac{7}{16}$	$3\frac{7}{16}$
Diameter of Foot Pulley—Inches	16	16	16	16	20	20	24	24	28
Approx. Weights—Lbs.									
Terminals§	708	752	802	855	1185	1600	2512	3002	4281
Conveyor per Ft. Centers§	13.0	14.0	18.5	20.3	24.9	35.0	47.2	55.2	63.2
Guide Idlers per Set (page 137)	*	*	*	*	25.8	30.5	35.0	......	....

*Guide Idlers for these Conveyors are included in weight of "Conveyor per Foot Centers."

†In no case should separate motor drives be less than 1½ to 2 Horse Power.

§Terminals comprise Head, Counter and Foot Shafts, Bearings, Set Collars, Gear and Pinion; Head and Foot Pulleys with Belt to extend halfway around pulleys. Conveyor per Foot Centers comprise carrying and return idlers with the necessary belt.

For Erection Dimensions of above Conveyors, see page 150.

Horizontal Standard Belt Conveyors 201 to 300 Feet Centers

For 50 Pound Materials Such as Coal

No. of Conveyor	325	326	327	328	330	333	336	428	429
Size of Material, Inches (page 129)									
Uniform Size or 70% to 80% of Unsized Material	2	2½	3	3½	4½	6	7½	9	10½
With Largest pieces not to exceed 10% of all	3	4	5	6	8	11	14	17	20
Capacity—(pages 129-130)									
Tons per Hour	36	48	70	86	143	254	404	601	862
Speed in Feet per Minute	225	225	260	260	300	340	375	410	450
Belt, "Century" (page 142)									
Width—Inches	14	16	18	20	24	30	36	42	48
Ply, with 1/16" Rubber Cover	4	4	4	4	5	5	6	6	7
Spacing of Carriers									
Three Pulley Troughing—Inches (page 134)	60	60	54	54	....	....	....	....	....
Five Pulley Troughing—Inches (page 135)	...	....	....	....	48	48	42	42	42
Side Hanging Returns, Feet (page 138)	10	10	10	10	10	10	10	10	10
Head Shaft. At discharge end									
Diameter of Shaft—Inches	1 15/16	1 15/16	1 15/16	2 7/16	2 15/16	2 15/16	3 7/16	3 15/16	4 7/16
Diameter of Pulley—Inches	20	20	20	20	24	24	30	30	36
Diameter of Gear—Inches	23.89	23.89	23.89	23.89	29.83	32.00	36.78	41.24	36.78
Pitch of Gear—Inches	1	1	1	1	1¼	1½	1¾	1¾	1¾
Face of Gear—Inches	2½	2½	2½	2½	3	4	5½	6	5½
Counter Shaft									
Diameter of Shaft—Inches	1 7/16	1 7/16	1 7/16	1 15/16	2 7/16	2 7/16	2 11/16	2 15/16	3 7/16
Rev. per Minute	202	202	234	234	240	241	225	257	226
Diameter of Pinion—Inches	5.12	5.12	5.12	5.12	6.01	7.22	7.86	8.42	7.86
Horse Power (page 131) at Counter Shaft for 300 feet Centers	2.7	3.6	5.4	6.6	9.9	14.4	23.1	31.8	42.6
Diameter of Foot Shaft—Inches	1 15/16	1 15/16	1 15/16	1 15/16	2 7/16	2 11/16	2 15/16	3 7/16	3 7/16
Diameter of Foot Pulley—Inches	16	16	16	16	20	20	24	24	28
Approx. Weight—Lbs.									
Terminals§	710	752	802	928	1459	1996	2965	3837	5029
Conveyor per Ft. Centers	13.0	14.0	15.8	18.9	29.2	35.0	46.8	53.9	61.7
Guide Idlers per Set. (page 137)	*	*	*	*	25.8	30.5	35.0	....	....

*Guide Idlers for these conveyors are included in weight of "Belt and Idlers per Foot Centers."

§Terminals comprise Head, Counter and Foot Shafts, Bearings, Set Collars, Gear and Pinion; Head and Foot Pulleys with Belt to extend halfway around pulleys. Conveyor per Foot Centers comprise carrying and return idlers with the necessary belt.

For Erection Dimensions of above Conveyors, see page 150.

Horizontal Standard Belt Conveyors 301 to 400 Feet Centers

For 50 Pound Materials Such as Coal

No. of Conveyor	337	338	339	340	342	345	348	430	431
Size of Material, Inches (page 129)									
Uniform or 70% to 80% of Size Unsized Material	2	$2\frac{1}{2}$	3	$3\frac{1}{2}$	$4\frac{1}{2}$	6	$7\frac{1}{2}$	9	$10\frac{1}{2}$
With Largest pieces not to exceed 10% of all	3	4	5	6	8	11	14	17	20
Capacity, (pages 129-130)									
Tons per Hour	36	48	70	86	143	254	404	601	862
Speed in Feet per Minute	225	225	260	260	300	340	375	410	450
Belt, "Century" (page 142)									
Width—Inches	14	16	18	20	24	30	36	42	48
Ply, with $\frac{1}{16}$" Rubber Cover	4	4	4	4	5	5	6	6	7
Spacing of Carriers									
Three Pulley Troughing—Inches (page 134)	60	60	54	54	....	....	....	....	....
Five Pulley Troughing—Inches (page 135)	....	....	....	....	48	48	42	42	42
Side Hanging Returns, Feet (page 138)	10	10	10	10	10	10	10	10	10
Head Shaft. At Discharge End									
Diameter of Shaft—Inches	$1\frac{13}{16}$	$2\frac{7}{16}$	$2\frac{7}{16}$	$2\frac{13}{16}$	$2\frac{15}{16}$	$3\frac{7}{16}$	$3\frac{15}{16}$	$4\frac{7}{16}$	$4\frac{15}{16}$
Diameter of Pulley—Inches	20	20	20	20	24	24	30	30	36
Diameter of Gear—Inches	23.89	23.89	29.83	29.83	32.00	36.78	36.78	36.78	36.78
Pitch of Gear—Inches	1	1	$1\frac{1}{4}$	$1\frac{1}{4}$	$1\frac{1}{2}$	$1\frac{3}{4}$	$1\frac{3}{4}$	$1\frac{3}{4}$	$1\frac{3}{4}$
Face of Gear—Inches	$2\frac{1}{2}$	$2\frac{1}{2}$	3	3	4	$5\frac{1}{2}$	$5\frac{1}{2}$	$5\frac{1}{2}$	$5\frac{1}{2}$
Counter Shaft									
Diameter of Shaft—Inches	$1\frac{7}{16}$	$1\frac{13}{16}$	$1\frac{15}{16}$	$2\frac{7}{16}$	$2\frac{7}{16}$	$2\frac{11}{16}$	$2\frac{15}{16}$	$3\frac{7}{16}$	$3\frac{15}{16}$
Rev. per Minute	202	202	250	250	215	254	226	245	226
Diameter of Pinion—Inches	5.12	5.12	6.01	6.01	7.22	7.86	7.86	7.86	7.86
Horse Power (page 131) At Counter Shaft for 400 Feet Centers	3.6	4.8	7.2	8.8	13.2	19.2	30.8	42.4	56.8
Diameter of Foot Shaft—Inches	$1\frac{15}{16}$	$1\frac{15}{16}$	$1\frac{15}{16}$	$2\frac{7}{16}$	$2\frac{11}{16}$	$2\frac{15}{16}$	$3\frac{7}{16}$	$3\frac{15}{16}$	$4\frac{7}{16}$
Diameter of Foot Pulley—Inches	16	16	16	16	20	20	24	24	28
Approx. Weights—Lbs.									
Terminals§	708	825	942	1199	1816	2438	3456	4244	5776
Conveyor per Ft. Centers §	13.2	14.0	18.3	20.0	29.4	35.2	47.0	54.0	61.8
Guide Idlers per Set. (page 137)	*	*	*	*	25.8	30.5	35.0	....	....

*Guide Idlers for these conveyors are included in weight of "Conveyor per Foot Centers."

§Terminals comprise Head, Counter and Foot Shafts, Bearings, Set Collars, Gear and Pinion; Head and Foot Pulleys with Belt to extend halfway around pulleys. Conveyor per Foot Centers comprise carrying and return idlers with the necessary belt.

For Erection Dimensions of above Conveyors, see page 151.

Horizontal Standard Belt Conveyors 401 to 500 Feet Centers

For 50 Pound Materials Such as Coal

No. of Conveyor	349	350	351	352	354	357	360	432	433
Size of Material, Inches (page 129)									
Uniform Size or 70% to 80% of Unsized Material	2	2½	3	3½	4½	6	7½	9	10½
With Largest Pieces not to exceed 10% of all	3	4	5	6	8	11	14	17	20
Capacity, (page 129-130)									
Tons per Hour	36	48	70	86	143	254	404	601	862
Speed in Feet per Minute	225	225	260	260	300	340	375	410	450
Belt, "Century" (page 142)									
Width—Inches	14	16	18	20	24	30	36	42	48
Ply, with $\frac{1}{16}$" Rubber Cover	4	4	4	4	5	5	6	6	7
Spacing of Carriers									
Three Pulley Troughing—Inches (page 134)	60	60	54	54	....	....	....	....	....
Five Pulley Troughing—Inches (page 135)	....	....	....	....	48	48	42	42	42
Side Hanging Returns, Feet (page 138)	10	10	10	10	10	10	10	10	10
Head Shaft. At Discharge End									
Diameter of Shaft—Inches	$2\frac{7}{16}$	$2\frac{15}{16}$	$2\frac{15}{16}$	$2\frac{15}{16}$	$2\frac{15}{16}$	$3\frac{7}{16}$	$4\frac{7}{16}$	$4\frac{7}{16}$	$4\frac{15}{16}$
Diameter of Pulley—Inches	20	20	20	20	24	24	30	30	36
Diameter of Gear—Inches	23.89	23.89	29.83	29.83	32.00	36.78	36.78	36.78	36.78
Pitch of Gear—Inches	1	1	1¼	1¼	1½	1¾	1¾	1¾	1¾
Face of Gear—Inches	2½	2½	3	3	4	5½	5½	5½	5½
Counter Shaft									
Diameter of Shaft—Inches	$1\frac{15}{16}$	$2\frac{7}{16}$	$2\frac{7}{16}$	$2\frac{7}{16}$	$2\frac{7}{16}$	$2\frac{11}{16}$	$3\frac{7}{16}$	$3\frac{7}{16}$	$3\frac{15}{16}$
Rev. per Minute	202	202	249	249	214	254	225	245	226
Diameter of Pinion—Inches	5.12	5.12	6.01	6.01	7.22	7.86	7.86	7.86	7.86
Horse Power (page 131) at Counter Shaft for 500 Feet Centers	4.5	6.0	9.0	11.0	16.5	24.0	38.5	53.0	71.0
Diameter of Foot Shaft—Inches	$1\frac{15}{16}$	$1\frac{15}{16}$	$2\frac{7}{16}$	$2\frac{7}{16}$	$2\frac{11}{16}$	$3\frac{7}{16}$	$3\frac{15}{16}$	$3\frac{15}{16}$	$4\frac{7}{16}$
Diameter of Foot Pulley—Inches	16	16	16	16	20	20	24	24	28
Approx. Weights—Lbs.									
Terminals§	786	974	1153	1208	1819	2657	3987	4248	5783
Conveyor per Ft. Centers§	13.4	14.1	18.3	20.1	29.2	35.2	46.9	53.9	61.7
Guide Idlers per Set, (page 137)	*	*	*	*	25.8	30.5	35.0	....	....

*Guide Idlers for these conveyors are included in weight of "Conveyor per Ft. Centers."

§Terminals comprise Head, Counter and Foot Shafts, Bearings, Set Collars, Gear and Pinion; Head and Foot Pulleys with Belt to extend halfway around pulleys. Conveyor per Foot Centers comprises carrying and return idlers with the necessary belt.

For Erection Dimensions of above Conveyors, see page 151.

Horizontal Standard Belt Conveyors 501 to 600 Feet Centers

For 50 Pound Materials Such as Coal

No. of Conveyor	361	362	363	364	366	369	372	434	435
Size of Material, Inches (page 129)									
Uniform Size or 70% to 80% of Unsized Material	2	2½	3	3½	4½	6	7½	9	10½
With Largest Pieces not to exceed 10% of all	3	4	5	6	8	11	14	17	20
Capacity, (pages 129-130)									
Tons per Hour	36	48	70	86	143	254	404	601	862
Speed in Feet per Minute	225	225	260	260	300	340	375	410	450
Belt, "Century" (page 142)									
Width—Inches	14	16	18	20	24	30	36	42	48
Ply, with $\frac{1}{16}''$ Rubber Cover	4	4	4	4	5	5	6	6	7
Spacing of Carriers									
Three Pulley Troughing—Inches (page 134)	60	60	54	54	....	....	....	....	....
Five Pulley Troughing—Inches (page 135)	----	----	----	----	48	48	42	42	42
Side Hanging Returns, Feet (page 138)	10	10	10	10	10	10	10	10	10
Head Shaft. At Discharge End									
Diameter of Shaft—Inches	$2\frac{7}{16}$	$2\frac{15}{16}$	$2\frac{15}{16}$	$2\frac{15}{16}$	$3\frac{7}{16}$	$3\frac{15}{16}$	$4\frac{7}{16}$	$4\frac{15}{16}$	$5\frac{7}{16}$
Diameter of Pulley—Inches	20	20	20	20	24	24	30	30	36
Diameter of Gear—Inches	23.89	29.83	29.83	32.00	36.78	36.78	36.78	36.78	48.41
Pitch of Gear—Inches	1	1¼	1¼	1½	1¾	1¾	1¾	1¾	2
Face of Gear—Inches	2½	3	3	4	5½	5½	5½	5½	6
Counter Shaft									
Diameter of Shaft—Inches	$1\frac{15}{16}$	$2\frac{7}{16}$	$2\frac{7}{16}$	$2\frac{7}{16}$	$2\frac{11}{16}$	$2\frac{15}{16}$	$3\frac{7}{16}$	$3\frac{15}{16}$	$4\frac{7}{16}$
Rev. per Minute	202	215	249	223	225	254	225	245	240
Diameter of Pinion—Inches	5.12	6.01	6.01	7.22	7.86	7.86	7.86	7.86	9.62
Horse Power (page 131) at Counter Shaft for 600 Feet Centers	5.4	7.2	10.8	13.2	19.8	28.8	46.2	63.6	85.2
Diameter of Foot Shaft—Inches	$1\frac{15}{16}$	$2\frac{7}{16}$	$2\frac{7}{16}$	$2\frac{11}{16}$	$2\frac{15}{16}$	$3\frac{7}{16}$	$3\frac{15}{16}$	$4\frac{7}{16}$	$4\frac{15}{16}$
Diameter of Foot Pulley—Inches	16	16	16	16	20	20	24	24	28
Approx. Weights—Lbs.									
Terminals§	787	1096	1152	1555	2248	2914	3997	4837	7577
Conveyor per Ft. Centers §	13.2	14.0	18.4	20.2	29.2	35.2	46.9	54.0	61.4
Guide Idlers per Set, (Page 137)	*	*	*	*	25.8	30.5	35.0		

*Guide Idlers for these conveyors are included in weight of "Conveyor per Ft. Centers."

§Terminals comprise Head, Counter and Foot Shafts, Bearings, Set Collars, Gear and Pinion; Head and Foot Pulleys with Belt to extend halfway around pulleys. Conveyor per Foot Centers comprises carrying and return idlers with the necessary belt.

For Erection Dimensions of above Conveyors, see page 151.

Horizontal Standard Belt Conveyors

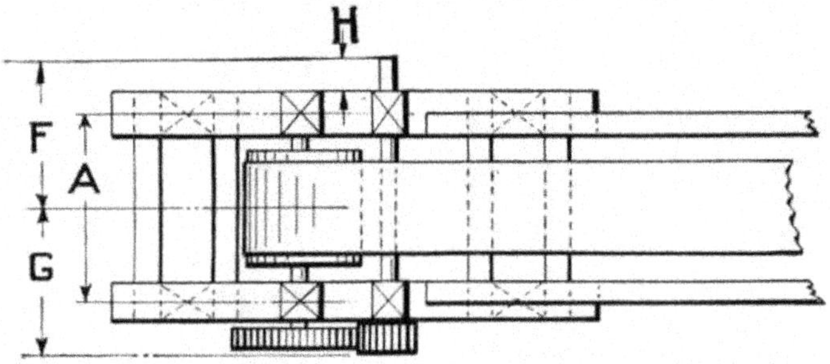

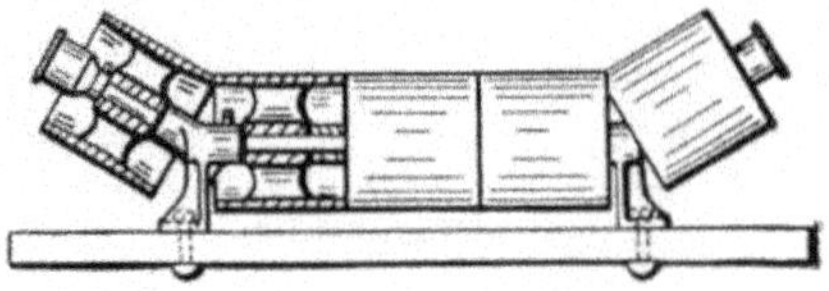

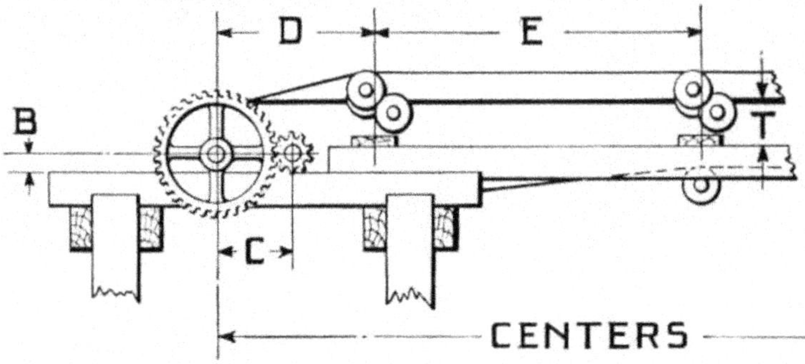

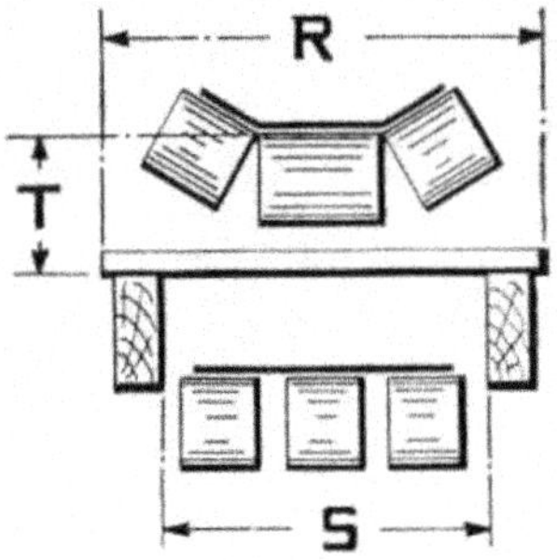

General Dimensions for 50 Pound Materials Such as Coal

Width of Belt In.	No. of Conveyor	A In.	B In.	C In.	D In.	E In.	F In.	G In.	H In.	J In.	M In.	R In.	S In.	T In.
0—100 FEET CENTERS.														
14	**301**	26	1 11/16	14½	30	60	21¼	20¼	6	2¼	11¾	24	20	7⅜
16	**302**	28	1 11/16	14½	30	60	22¼	21¼	6	2¼	11¾	26	22	7⅜
18	**303**	30	1 11/16	14½	36	54	23¼	22¼	6	2¼	11¾	30	24	8⅜
20	**304**	32	1 11/16	14½	36	54	24¼	23¼	6	2¼	11¾	32	26	8⅜
24	**306**	36	3⅛	14½	36	48	27	27	6	2¾	12	36	30	8⅞
30	**309**	44	3⅛	14½	36	48	31	31	6	2¾	12	44	38	8⅞
36	**312**	50	3⅝	18	36	42	34¾	35¾	6	3⅛	15	50	44	8⅞
42	**421**	56	3⅝	18	36	42	37¾	38¾	6	4	11	56	50	8⅞
48	**422**	62	4	23¾	36	42	42⅛	43½	7	4	10½	62	56	8⅞
101—200 FEET CENTERS.														
14	**313**	26	1 11/16	14½	30	60	21¼	20¼	6	2¾	12	24	20	7⅜
16	**314**	28	1 11/16	14½	30	60	22¼	21¼	6	2¾	12	26	22	7⅜
18	**315**	30	1 11/16	14½	36	54	23¼	22¼	6	2¾	12	30	24	8⅜
20	**316**	32	1 11/16	14½	36	54	24¼	23¼	6	2¾	12	32	26	8⅜
24	**318**	36	3⅛	14½	36	48	27	27	6	2¾	12	36	30	8⅞
30	**321**	44	3⅝	18	36	48	31¾	32¾	6	3⅜	15	44	38	8⅞
36	**324**	50	4	19⅝	36	42	36⅛	37½	7	4	10½	50	44	8⅞
42	**426**	56	4	22 5/16	36	42	39⅛	40½	7	4	10½	56	50	8⅞
48	**427**	62	4⅝	22 5/16	36	42	43½	45¼	8	5	15¼	62	56	8⅞
201—300 FEET CENTERS.														
14	**325**	26	1 11/16	14½	30	60	21¼	20¼	6	2¾	18	24	20	7⅜
16	**326**	28	1 11/16	14½	30	60	22¼	21¼	6	2¾	18	26	22	7⅜
18	**327**	30	1 11/16	14½	36	54	23¼	22¼	6	2¾	18	30	24	8⅜
20	**328**	32	3⅛	14½	36	54	25	25	6	2¾	18	32	26	8⅜
24	**330**	36	3⅜	18	36	48	27¾	28¾	6	3⅛	20	36	30	8⅞
30	**333**	44	3⅛	19⅝	36	48	31¾	32¾	6	4	22¾	44	38	8⅞
36	**336**	50	4	22 5/16	36	42	36⅛	37½	7	4	22¼	50	44	8⅞
42	**428**	56	4⅝	24⅞	36	42	40½	42¼	8	5⅛	29¾	56	50	8⅞
48	**429**	62	5⅜	22 5/16	36	42	45¼	47	9	5¼	35¾	62	56	8⅞

Horizontal Standard Belt Conveyors

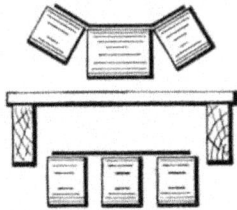

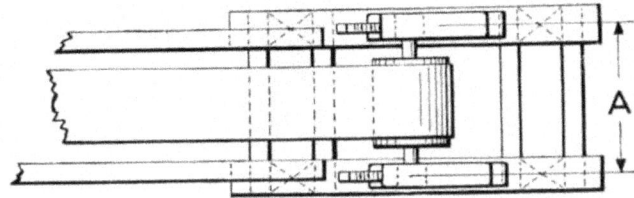

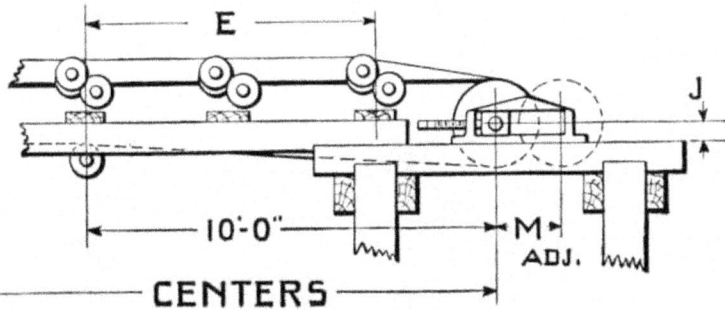

General Dimensions for 50 Pound Materials Such as Coal

Width of Belt In.	No. of Conveyor	A In.	B In.	C In.	D In.	E In.	F In.	G In.	H In.	J In.	M In.	R In.	S In.	T In.
301—400 FEET CENTERS.														
14	337	26	1 11/16	14½	30	60	21¼	20¼	6	2¾	18	24	20	7⅜
16	338	28	3⅛	14½	30	60	23	23	6	2¾	18	26	22	7⅜
18	339	30	3⅛	18	36	54	24	24	6	2¾	18	30	24	8⅜
20	340	32	3⅝	18	36	54	25¾	26¾	6	3⅛	20	32	26	8⅜
24	342	38	3⅝	19⅝	36	48	28¾	29¾	6	4	22¾	36	30	8⅞
30	345	44	4	22 5/16	36	48	33⅛	34½	7	4	22¼	44	38	8⅞
36	348	50	4⅝	22 5/16	36	42	37½	39¼	8	5⅛	29¾	50	44	8⅞
42	430	58	5⅜	22 5/16	36	42	43¼	45	9	5¼	35¾	56	50	8⅞
48	431	64	5⅝	22 5/16	36	42	48	49¾	10	6 1/16	24¼	62	56	8⅞
401—500 FEET CENTERS.														
14	349	28	3⅛	14½	30	60	23	23	6	2¾	18	24	20	7⅜
16	350	30	3⅝	14½	30	60	24¾	25¾	6	2¾	18	26	22	7⅜
18	351	32	3⅝	18	36	54	25¾	26¾	6	3⅛	20	30	24	8⅜
20	352	34	3⅝	18	36	54	26¾	27¾	6	3⅛	20	32	26	8⅜
24	354	38	3⅝	19⅝	36	48	28¾	29¾	6	4	22¾	36	30	8⅞
30	357	44	4	22 5/16	36	48	33⅛	34½	7	5⅛	29¾	44	38	8⅞
36	360	52	5⅜	22 5/16	36	42	40¼	42	9	5¼	35¾	50	44	8⅞
42	432	58	5⅜	22 5/16	36	42	43¼	45	9	5¼	35¾	56	50	8⅞
48	433	64	5⅝	22 5/16	36	42	48½	49¼	10	6 1/16	24¼	62	56	8⅞
501—600 FEET CENTERS.														
14	361	28	3⅛	14½	30	60	23	23	6	2¾	18	24	20	7⅜
16	362	30	3⅝	18	30	60	24¾	25¾	6	3⅛	20	26	22	7⅜
18	363	32	3⅝	18	36	54	25¾	26¾	6	3⅛	20	30	24	8⅜
20	364	34	3⅝	19⅝	36	54	26¾	27¾	6	4	22¾	32	26	8⅜
24	366	38	4	22 5/16	36	48	30⅛	31½	7	4	22¾	36	30	8⅞
30	369	44	4⅝	22 5/16	36	48	34½	36¼	8	5⅛	29¾	44	38	8⅞
36	372	52	5⅜	22 5/16	36	42	40¼	42	9	5¾	35¾	50	44	8⅞
42	434	58	5⅝	22 5/16	36	42	45½	46¼	16	6 1/16	24¼	56	50	8⅞
48	435	64	6¼	29	36	42	49¾	51½	11	6¾	37⅝	62	56	8⅞

Steel Apron Conveyors

Section

8

Jeffrey Standard Steel Apron Conveyor operating from under track hopper as shown in lower left hand illustration, for carrying coal to crusher.

Another view of the same Steel Apron Conveyor showing the cleats placed at intervals to enable it to carry up the incline.

Jeffrey Steel Apron Conveyor handling coal from under track hopper and crusher to storage pit.

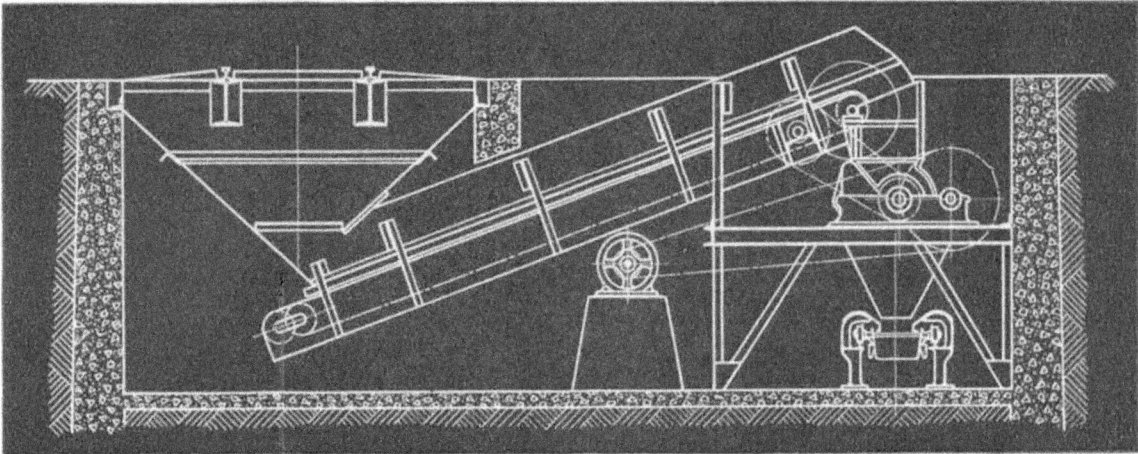

A typical layout of Jeffrey Standard Steel Apron Feeder Conveyor. The Conveyor receives coal through track-hopper and discharges it into Crusher. This arrangement is very simple and eliminates deep excavations. Other arrangements illustrated on pages 86 and 87.

YEARS of hard service have proven Jeffrey Standard Steel Apron Conveyors to be particularly adapted to Boiler House service, as feeder to Crusher or intermediate conveyor between crusher and main conveyor.

The steel flights are beaded on the edge so as to overlap, forming a continuous moving surface. They are mounted upon two strands of roller chain, and are capable of carrying up quite steep inclines as well as along the horizontal.

Sectional View of Steel Apron Conveyor, with Jeffrey popular No. 126-C Malleable Roller Chain. Jeffrey Steel Apron Conveyors are made in various sizes to suit capacity requirements and special conditions.

Receiving end of a Jeffrey Standard Steel Apron Conveyor with steel chains, operating from under track hopper, carrying coal to Crusher.

Jeffrey Standard Steel Apron Conveyor delivering coal from track hopper to storage.

Steel Apron Feeder carrying coal from track hopper to Crusher

A Dependable Feeder

JEFFREY Standard Steel Apron Conveyor, equipped with skirtboards, serving as a feeder conveyor delivering run-of-mine coal from track hopper to a Jeffrey 30″ x 30″ Single Roll Crusher. The Apron is 30″ wide, mounted on 126-C Malleable Roller Chain, page 162. From beneath the crusher a scraper conveyor carries the crushed coal up an incline into storage bins from where it is spouted into the gas producers of a large glass factory.

Some Important Points to Know Regarding Apron Conveyors

THE Steel Apron consists of two strands of roller chain between which are bolted double beaded steel flights. These flights as shown in the line illustration, are so made that they always overlap thus making a tight apron. The flights are provided with steel ends which in connection with the apron form a continuous moving trough. This type of conveyor is intended for conveying any kind of loose bulk materials which are not of a sticky nature such as coal, ores, stone, gravel, cullets, steel scrap, etc.

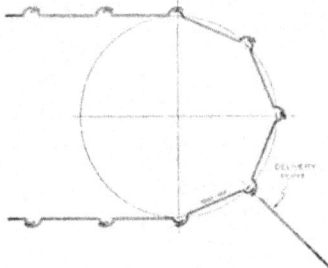

"THE DOUBLE BEADED APRON" is the most popular type of steel conveyor and is practically leakage proof in the carrying of non-sticky coarse materials of all kinds. Furnished with retaining ends for loose materials and without ends for merchandise.

Steel Apron Used as Conveyor or Feeder

The steel apron conveyor, when used in connection with steel skirt boards as illustrated on opposite page, makes a most satisfactory feeder, insuring a continuous, uniform flow of material.

The Standard Conveyors are divided into four sets based on the length of conveyor. The first set includes Conveyors from 0 to 25 feet centers, the second from 26 to 50 feet, the third from 51 to 75 feet, and the fourth from 76 to 100 feet centers.

They may be installed on the horizontal, incline or a combination of incline and horizontal. The corrugated effect of the double beaded flight serves as a check against the flow of material. The angle of inclination, however, should not exceed 30 degrees for such materials as coal, ores, stone, etc.

Double Beaded Apron Practically Leakage Proof

Jeffrey Steel Apron Conveyors are practically noiseless in operation, require but little power and as the material is at all times carried by the conveyor, the item of breakage is negligible which makes them ideal for friable materials.

Because of their flexibility and wide range of sizes the Standard Steel Apron Conveyors will meet the requirements of almost any conveying problem which might present itself.

Things to Note When Ordering your Apron Conveyor

Each Conveyor Complete in Itself

THE specifications of Jeffrey Standard Apron Conveyors shown on the following pages cover all the necessary machinery parts for a complete conveyor. Wearing strips with wood screws are furnished with each conveyor for both the carrying and return runways.

The supports may be of steel or wood construction. If wood is used, any good carpenter or millwright can erect suitable supports and install the machinery parts by following the general erection drawings on Pages 161 to 169.

Erecting a Steel Apron Conveyor

In the erection of a Steel Apron Conveyor it is essential that the bottoms of the flights be on the center line of the chain, also that the larger of the two beads and the pointed part of the steel ends face in the same direction as the chain travels. See illustration above.

Walkways should be provided for all conveyors where the chain and flights would otherwise be inaccessible.

How to Figure Shipping Weights

Shipping Weights of any of the Standard Apron Conveyors may be readily figured by referring to Tables of Specifications under the heading of Approximate Shipping Weight, Apron Complete per Foot Centers and multiply the value there given by the centers of the required conveyor. To this product add the Weight of "Terminals Complete".

The Terminals comprise Head Shaft with sprockets, bearings and larger gear; Counter Shaft with extension for purchasers drive pulley, bearings, set collar and pinion; Foot Shaft with takeup bearings, set collars and sprockets; also sufficient chain and flights to extend half way around both the head and foot sprockets. "Conveyor per Foot Centers" consists of the necessary wearing strip, chain and flights to make up one foot of both the carrying and return strands.

When ordering or referring to Standard Apron Conveyors, always give Conveyor Number and its Centers

Capacities of Jeffrey Standard Steel Apron Conveyors

THE Standard Steel Apron Conveyors are divided into two classes, those handling material weighing not in excess of 50 lbs. per cubic foot and those handling material weighing not over 100 lbs. per cubic foot. The capacities of Steel Apron Conveyors handling loose materials are based upon the conveyors being loaded uniformly throughout and traveling at the rate of 100 feet per

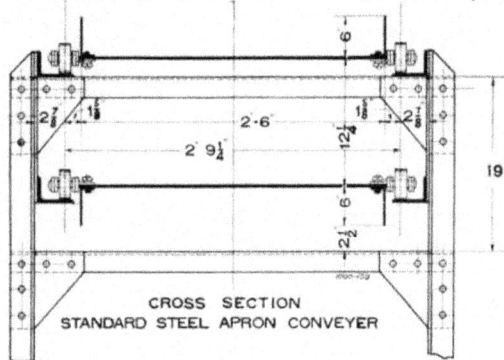

CROSS SECTION
STANDARD STEEL APRON CONVEYER

minute. The values given in the Tables of Specifications, pages 160 to 168, are 80% of the level full capacity, this percentage having been found to be a very good average. The capacities vary from 45 to 400 tons per hour for the conveyors handling 50 lb. materials, and from 120 to 800 tons per hour for 100 lb. materials.

Determine First the Size of Material

The size of material handled may vary from very fine to 24″ cubes, the "Width of Apron" being the governing factor. It will be noted in the Specification Tables on pages 160 to 168, under "Size of Material" that there are two divisions, one giving the "Average Size of Material to be handled" and the other division showing the "Maximum Size Pieces" which the conveyor will handle, with the notation that the amount of such pieces should not exceed 10% of all the material. If this percentage is exceeded the next larger size conveyor should be specified irrespective of the capacity.

How to find Capacity for Different Materials

If material of a lesser weight is to be handled than that indicated the capacity must be reduced in direct proportion. For example it is desired to convey wood chips at the rate of 30 tons per hour, the chips weighing approximately 20 lbs. per cubic foot. The Standard Conveyors will handle just as big a volume of chips at 20 lbs. per cubic foot as they will of material weighing 50 lbs. per cubic foot, but by weight a conveyor with a listed capacity of a certain number of tons per hour of 50 lb. material will handle only $\frac{20}{50}$ of that amount when handling material

weighing 20 lb. per cubic foot. From the above it is evident that to handle 30 tons per hour of 20 lb. material it will be necessary to select a conveyor from the tables with a capacity equal to $\frac{50}{20}$ or $2\frac{1}{2}$ times the required capacity. On the other hand if the material to be handled is of a greater weight than that for which the conveyor was figured some sort of feeding device should be provided to insure against overloading the conveyor.

Standard Conveyor Used as a Feeder

When a Standard Apron Conveyor is used as an Apron Feeder, the loaded condition is just the reverse of that when it is used simply as a conveyor. As will be noted from the line cut the Apron Feeder is fitted with steel skirt boards usually about 24″ high, which permit of carrying a deep load. To insure a uniform depth of load the conveyor is operated at a slow speed. The average depth of the load with 24″ skirt boards is about 18″, which is three times as much as a standard conveyor will carry with 6″ high ends. It is evident from the above that if a Standard

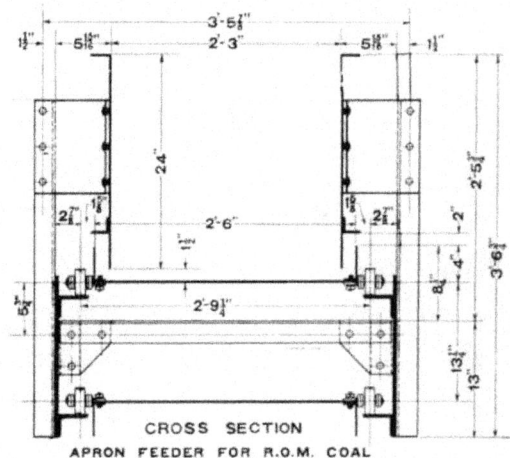

CROSS SECTION
APRON FEEDER FOR R.O.M. COAL

Conveyor is to be used as a feeder a conveyor must be selected from the tables which is approximately three times as long as the feeder conveyor. Under these conditions the total load on the two conveyors is about the same and the shafting and gearing required by the long conveyor under normal loading conditions are in keeping with the requirements of a conveyor, one third as long, serving as a feeder conveyor. A safe rule to follow when a conveyor is to serve as a feeder with skirt boards is to select a conveyor from the tables which is three times the length of the feeder conveyor, and specify the centers to suit the requirements. The capacity of a feeder conveyor may be varied by increasing or decreasing the rate of travel.

Determining Horse-Power for Jeffrey Apron Conveyors

JEFFREY Standard Steel Apron Conveyors are very economical in matter of power consumption as is evidenced by the horse-power ratings of the various conveyors in the specification tables, pages 160 to 168. These listed ratings represent the power required at the counter shaft to operate a horizontal conveyor, of the maximum centers given at the top of each table, at a speed of 100 ft. per minute.

Horizontal Conveyors

The speeds specified above are about the maximum for good service and proper loading of the conveyors and should not be increased but may be decreased if desired in which case the capacity would be decreased in direct proportion. This is also true of the horsepower which is directly proportional to the speed and varies accordingly. For example: Conveyor No. 2551, page 168, for maximum centers of 100 feet requires 12 horsepower to operate at the specified speed of 100 ft. per minute. As the horsepower of any conveyor is equal to $\dfrac{\text{Load} \times \text{speed in ft. per min.}}{33000}$ it is plain to be seen that if the speed of conveyor No. 2551 be reduced one fourth or to 75 ft. per minute the horsepower also is reduced one-fourth.

In the formula above the load as well as the speed is a function of the horsepower. As the load is determined by the length or "Centers" of a conveyor it is evident that the horsepower of any conveyor is dependent upon its length and varies directly as the length is increased or decreased. This is brought out very clearly in the tables of specifications. For example turn to page 168 and compare the horsepower ratings of Conveyors No. 2275, 2367, 2459 and 2551. These four conveyors are identical except in their lengths, there being a difference of 25 feet between each. Note that the horsepower of Conveyor No. 2367 is twice that of 2275 also that the centers of 2367 is twice that of 2275. Conveyor 2459 which is three times as long as 2275 requires three times as much power and so on. From the above it is apparent that if the centers of a required conveyor are less than that listed in the tables the horsepower may be reduced accordingly.

Incline Conveyors

While the horsepower ratings in the tables are for horizontal conveyors any one of the conveyors may be installed on an incline providing the angle of inclination does not exceed 30 degrees from the horizontal.

Nominally, the horsepower required for an inclined conveyor is the same as that required for a horizontal conveyor whose centers are equal to that of the inclined conveyor plus three feet for each foot of rise. For example a conveyor of 60 feet inclined centers which has a rise of 10 ft. requires the same horsepower as a horizontal conveyor of 60 plus 3 times 10 or 90 feet centers.

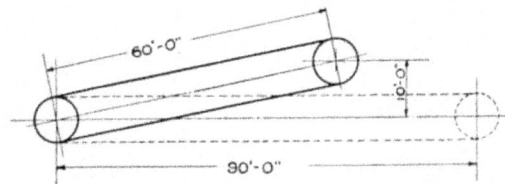

All of the Standard Apron Conveyors are provided with one set of cast teeth gears and a countershaft which has a keyseated extension to receive purchasers drive pulley or sprocket. See table below. If the speeds, given in the tables of specifications, for the countershafts are not sufficient to connect with the source of power, an additional set of gears with a second countershaft should be provided. The diameter of this second countershaft may be safely taken as $\frac{1}{2}''$ less than the first countershaft with $1\frac{3}{16}''$ Diameter as a minimum.

Table of Countershaft Extensions for Drive Pulleys

Diameter of countershaft	Extension	Size of Keyway	
		Width	Depth
$1\frac{3}{16}''$	$6''$	$\frac{5}{16}''$	$\frac{5}{32}''$
$1\frac{7}{16}''$	$6''$	$\frac{3}{8}''$	$\frac{3}{16}''$
$1\frac{15}{16}''$	$6''$	$\frac{1}{2}''$	$\frac{1}{4}''$
$2\frac{7}{16}''$	$6''$	$\frac{5}{8}''$	$\frac{5}{16}''$
$2\frac{11}{16}''$	$7''$	$\frac{11}{16}''$	$\frac{11}{32}''$
$2\frac{15}{16}''$	$8''$	$\frac{3}{4}''$	$\frac{3}{8}''$

If the operating conditions of a conveyor call for frequent stopping and starting under full load the horsepower listed in the tables should be increased 40%.

For separate motor drives use a 2 horsepower motor for all conveyors requiring less than 2 horsepower and a 3 horsepower motor for conveyors requiring between 2 and 3 horsepower.

Specifications of Jeffrey Standard Steel Apron Conveyors Using No. 14½ Malleable Roller Chain

Material Weighing Approximately 50 Pounds per Cubic Foot

Length of Conveyor	0 to 25 ft. Centers				26 to 50 ft. Centers				51 to 75 ft. Centers				76 to 100 ft. Centers			
No. of Conveyor	2192	2215	2238	2261	2284	2307	2330	2353	2376	2399	2422	2445	2468	2491	2514	2537
Size of Material																
Average size of Material to be handled	3	4	7	9	3	4	7	9	3	4	7	9	3	4	7	9
Max. Size not to exceed 10% of Whole	6	8	14	18	6	8	14	18	6	8	14	18	6	8	14	18
Load in Lbs. per Foot on Conveyor	20	27	34	40	20	27	34	40	20	27	34	40	20	27	34	40
Capacity—Tons per Hour	60	80	101	120	60	80	101	120	60	80	101	120	60	80	101	120
Width of Apron	18	24	30	36	18	24	30	36	18	24	30	36	18	24	30	36
Horse Power at Center Shaft	.78	1.0	1.2	1.4	1.6	1.9	2.3	2.7	2.3	2.9	3.5	4	3.1	3.9	4.7	5.3
Head Shaft																
Diameter, Inches	1 7/16	1 7/16	1 7/16	1 11/16	1 11/16	1 11/16	1 11/16	1 11/16	1 11/16	1 11/16	1 11/16	1 11/16	1 11/16	1 11/16	2 7/16	2 7/16
Rev. per Minute	38	38	38	38	38	38	38	38	38	38	38	38	38	38	38	38
Size Sprockets, Inches	10½	10½	10½	10½	10½	10½	10½	10½	10½	10½	10½	10½	10½	10½	10½	10½
Diameter of Gear	17.84	17.84	17.84	17.84	17.84	17.84	17.84	17.84	17.84	17.84	19.91	19.91	19.91	19.91	19.91	19.91
Pitch of Gear	1	1	1	1	1	1	1	1	1	1	1¼	1¼	1¼	1¼	1¼	1¼
Face of Gear	2	2	2	2	2	2	2	2	2	2	3	3	3	3	3	3
Counter Shaft																
Diameter, Inches	1 3/16	1 3/16	1 3/16	1 3/16	1 3/16	1 3/16	1 3/16	1 3/16	1 3/16	1 3/16	1 7/16	1 7/16	1 7/16	1 7/16	1 11/16	1 11/16
Rev. per Minute	133	133	133	133	133	133	133	133	133	133	127	127	127	127	127	127
Diameter of Pinion	5.12	5.12	5.12	5.12	5.12	5.12	5.12	5.12	5.12	5.12	6.01	6.01	6.01	6.01	6.01	6.01
Face of Pinion	2¾	2¾	2¾	2¾	2¾	2¾	2¾	2¾	2¾	2¾	3¼	3¼	3¼	3¼	3¼	3¼
Foot Shaft																
Diameter, Inches	1 7/16	1 7/16	1 7/16	1 7/16	1 7/16	1 7/16	1 7/16	1 7/16	1 7/16	1 7/16	1 7/16	1 7/16	1 7/16	1 7/16	1 11/16	1 11/16
Size Sprockets, Inches	10½	10½	10½	10½	10½	10½	10½	10½	10½	10½	10½	10½	10½	10½	10½	10½
Approx. Shipping Weight—Lbs.																
Terminals, Complete	405	420	435	505	455	470	490	505	455	470	545	560	510	525	710	730
Apron Complete per Foot Centers	38	44	52	56	38	44	52	56	38	44	52	56	38	44	52	56

Material Weighing Approximately 100 Pounds per Cubic Foot

Length of Conveyor	0 to 25 ft. Centers				26 to 50 ft. Centers				51 to 75 ft. Centers				76 to 100 ft. Centers			
No. of Conveyor	2557	2577	2597	2617	2637	2657	2677	2697	2717	2737	2757	2777	2797	2817	2837	2857
Size of Material																
Average size of Material to be handled	3	4	7	9	3	4	7	9	3	4	7	9	3	4	7	9
Max. Size not to exceed 10% of Whole	6	8	14	18	6	8	14	18	6	8	14	18	6	8	14	18
Load in Lbs. per Foot on Conveyor	40	54	68	80	40	54	68	80	40	54	68	80	40	54	68	80
Capacity—Tons per Hour	120	160	202	240	120	160	202	240	120	160	202	240	120	160	202	240
Width of Apron	18	24	30	36	18	24	30	36	18	24	30	36	18	24	30	36
Horse Power at Center Shaft	1.1	1.3	1.6	1.9	2.2	2.7	3.2	3.7	3.2	4.0	4.8	5.6	4.2	5.3	6.4	7.5
Head Shaft																
Diameter, Inches	1 7/16	1 11/16	1 11/16	1 11/16	1 11/16	1 11/16	1 11/16	1 11/16	1 11/16	1 11/16	1 11/16	2 7/16	1 11/16	1 11/16	2 7/16	2 7/16
Rev. per Minute	38	38	38	38	38	38	38	38	38	38	38	38	38	38	38	38
Size Sprockets, Inches	10½	10½	10½	10½	10½	10½	10½	10½	10½	10½	10½	10½	10½	10½	10½	10½
Diameter of Gear	17.84	17.84	17.84	17.84	17.84	17.84	19.91	19.91	19.91	19.91	19.91	29.83	19.91	19.91	19.91	29.83
Pitch of Gear	1	1	1	1	1	1	1¼	1¼	1¼	1¼	1¼	1¼	1¼	1¼	1¼	1¼
Face of Gear	2	2	2	2	2	2	3	3	3	3	3	3	3	3	3	3
Counter Shaft																
Diameter, Inches	1 3/16	1 3/16	1 3/16	1 3/16	1 3/16	1 3/16	1 7/16	1 7/16	1 7/16	1 7/16	1 7/16	1 11/16	1 7/16	1 7/16	1 11/16	1 11/16
Rev. per Minute	133	133	133	133	133	133	127	127	127	127	127	190	127	127	127	190
Diameter of Pinion	5.12	5.12	5.12	5.12	5.12	5.12	6.01	6.01	6.01	6.01	6.01	6.01	6.01	6.01	6.01	6.01
Face of Pinion	2¾	2¾	2¾	2¾	2¾	2¾	3¼	3¼	3¼	3¼	3¼	3¼	3¼	3¼	3¼	3¼
Foot Shaft																
Diameter, Inches	1 7/16	1 7/16	1 7/16	1 7/16	1 7/16	1 7/16	1 7/16	1 7/16	1 7/16	1 7/16	1 7/16	1 11/16	1 7/16	1 7/16	1 11/16	1 11/16
Size Sprockets, Inches	10½	10½	10½	10½	10½	10½	10½	10½	10½	10½	10½	10½	10½	10½	10½	10½
Approx. Shipping Weight—Lbs.																
Terminals, Complete	405	470	495	505	455	470	545	560	510	525	545	820	510	525	710	820
Apron Complete per Foot Centers	38	44	52	56	38	44	52	56	38	44	52	56	38	44	52	56

For Erection Dimensions of the above Conveyors, see opposite page.

General Dimensions of Jeffrey Standard Steel Apron Conveyors Using No. 14½ Malleable Roller Chain

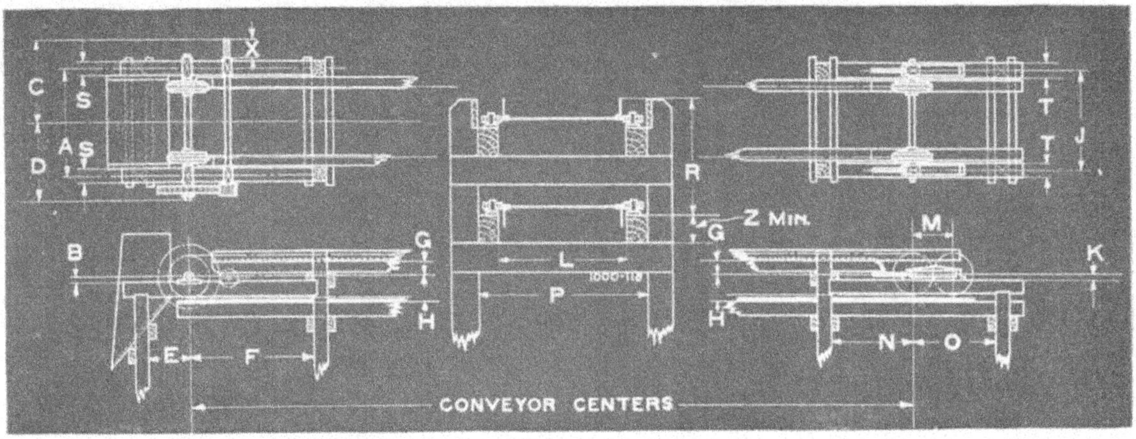

For Material Weighing 50 Pounds per Cubic Foot. Dimensions in Inches

Length of Conveyor	0 to 25 ft. Centers				26 to 50 ft. Centers				51 to 75 ft. Centers				76 to 100 ft. Centers			
No. of Conveyor	2192	2215	2238	2261	2284	2307	2330	2353	2376	2399	2422	2445	2468	2491	2514	2537
A	29½	35½	41½	48½	30½	36½	42½	48½	30½	36½	42½	48½	30½	36½	44¾	50¾
B	15/16	15/16	15/16	2¼	2¼	2¼	2¼	2¼	2¼	2¼	2¼	2¼	2¼	2¼	2⅝	2⅝
C	22⅝	25⅝	28⅝	32½	23½	26½	29½	32½	23½	26½	29½	32½	23½	26½	31⅜	34⅜
D	20¼	23¼	26¼	31½	22½	25½	28½	31½	22½	25½	28½	31½	22½	25½	31⅛	34⅛
E	12	12	12	12	12	12	12	12	12	12	12	12	12	12	12	12
F	25	25	25	25	25	25	25	25	25	25	25	25	25	25	25	25
G	3⅝	3⅝	3⅝	3⅝	3⅝	3⅝	3⅝	3⅝	3⅝	3⅝	3⅝	3⅝	3⅝	3⅝	3⅝	3⅝
H	7½	7½	7½	7½	7½	7½	7½	7½	7½	7½	7½	7½	7½	7½	7½	7½
J	29½	35½	41½	47½	29½	35½	41½	47½	29½	35½	41½	47½	29½	35½	42	48
K	2 9/32	2 9/32	2 9/32	2 9/32	2 9/32	2 9/32	2 9/32	2 9/32	2 9/32	2 9/32	2 9/32	2 9/32	2 9/32	2 9/32	2 25/32	2 25/32
L	19¼	25¼	31¼	37¼	19¼	25¼	31¼	37¼	19¼	25¼	31¼	37¼	19¼	25¼	31¼	37¼
M	11¾	11¾	11¾	11¾	11¾	11¾	11¾	11¾	11¾	11¾	11¾	11¾	11¾	11¾	12	12
N	23	23	23	23	23	23	23	23	23	23	23	23	23	23	25	25
O	24	24	24	24	24	24	24	24	24	24	24	24	24	24	24	24
P	25¼	31¼	37¼	43¼	25¼	31¼	37¼	43¼	25¼	31¼	37¼	43¼	25¼	31¼	37¼	43¼
R	16⅜	16⅜	16⅜	16⅜	16⅜	16⅜	16⅜	16⅜	16⅜	16⅜	16⅜	16⅜	16⅜	16⅜	16⅜	16⅜
S	4	4	4	4	4	4	4	4	4	4	4	4	4	4	6	6
T	4	4	4	4	4	4	4	4	4	4	4	4	4	4	4	4
X	6	6	6	6	6	6	6	6	6	6	6	6	6	6	6	6
Z	6	6	6	6	6	6	6	6	6	6	6	6	6	6	6	6

For Material Weighing 100 Pounds per Cubic Foot. Dimensions in Inches

Length of Conveyor	0 to 25 ft. Centers				26 to 50 ft. Centers				51 to 75 ft. Centers				76 to 100 ft. Centers			
No. of Conveyor	2557	2577	2597	2617	2637	2657	2677	2697	2717	2737	2757	2777	2797	2817	2837	2857
A	29½	36½	42½	48½	30½	36½	42½	48½	30½	36½	42½	50¾	30½	36½	44¾	50¾
B	15/16	2¼	2¼	2¼	2¼	2¼	2¼	2¼	2¼	2¼	2¼	2⅝	2¼	2¼	2⅝	2⅝
C	22⅝	26½	29½	32½	23½	26½	29½	32½	23½	26½	29½	34⅜	23½	26½	31⅜	34⅜
D	20¼	25½	28½	31½	22½	25½	28½	31½	22½	25½	28½	34⅛	22½	25½	31⅛	34⅛
E	12	12	12	12	12	12	12	12	12	12	12	12	12	12	12	12
F	25	25	25	25	25	25	25	25	25	25	25	30	25	25	25	30
G	3⅝	3⅝	3⅝	3⅝	3⅝	3⅝	3⅝	3⅝	3⅝	3⅝	3⅝	3⅝	3⅝	3⅝	3⅝	3⅝
H	7½	7½	7½	7½	7½	7½	7½	7½	7½	7½	7½	7½	7½	7½	7½	7½
J	29½	35½	41½	47½	29½	35½	41½	47½	29½	35½	41½	48	29½	35½	42	48
K	2 9/32	2 9/32	2 9/32	2 9/32	2 9/32	2 9/32	2 9/32	2 9/32	2 9/32	2 9/32	2 9/32	2 25/32	2 9/32	2 9/32	2 25/32	2 25/32
L	19¼	25¼	31¼	37¼	19¼	25¼	31¼	37¼	19¼	25¼	31¼	37¼	19¼	25¼	31¼	37¼
M	11¾	11¾	11¾	11¾	11¾	11¾	11¾	11¾	11¾	11¾	11¾	12	11¾	11¾	12	12
N	23	23	23	23	23	23	23	23	23	23	23	25	23	23	25	25
O	24	24	24	24	24	24	24	24	24	24	24	24	24	24	24	24
P	25¼	31¼	37¼	43¼	25¼	31¼	37¼	43¼	25¼	31¼	37¼	43¼	25¼	31¼	37¼	43¼
R	16⅜	16⅜	16⅜	16⅜	16⅜	16⅜	16⅜	16⅜	16⅜	16⅜	16⅜	16⅜	16⅜	16⅜	16⅜	16⅜
S	4	4	4	4	4	4	4	4	4	4	4	6	4	4	6	6
T	4	4	4	4	4	4	4	4	4	4	4	4	4	4	4	4
X	6	6	6	6	6	6	6	6	6	6	6	6	6	6	6	6
Z	6	6	6	6	6	6	6	6	6	6	6	6	6	6	6	6

Specifications of Jeffrey Standard Steel Apron Conveyors Using No. 126-C Malleable Roller Chain

Material Weighing Approximately 50 Pounds per Cubic Foot

Length of Conveyor	0 to 25 ft. Centers					26 to 50 ft. Centers					51 to 75 ft. Centers					76 to 100 ft. Centers				
No. of Conveyor	2197	2198	2221	2244	2267	2289	2290	2313	2336	2359	2381	2382	2405	2428	2451	2473	2474	2497	2520	2543
Size of Material																				
Average size to be handled	3½	4	7	9	14	3½	4	7	9	14	3½	4	7	9	14	3½	4	7	9	14
Max. Size not to exceed 10% of Whole	7	8	14	18	24	7	8	14	18	24	7	8	14	18	24	7	8	14	18	24
Load Lbs. per Ft. on Conveyor	30	40	50	60	80	30	40	50	60	80	30	40	50	60	80	30	40	50	60	80
Capacity—Tons per Hr.	100	120	149	180	240	100	120	149	180	240	100	120	149	180	240	100	120	149	180	240
Width of Apron	18	24	30	36	48	18	24	30	36	48	18	24	30	36	48	18	24	30	36	48
H. P. at Ctr. Shaft	1.5	1.8	2.0	2.4	2.9	3.0	3.6	4.2	4.8	5.9	4.5	5.3	6.2	7.1	8.8	6.0	7.2	8.3	9.5	11.8
Head Shaft																				
Diameter, Inches	1 15/16	1 15/16	1 15/16	1 15/16	1 15/16	1 15/16	1 15/16	2 7/16	2 7/16	2 7/16	2 7/16	2 7/16	2 7/16	2 15/16	2 15/16	2 15/16	2 15/16	2 15/16	2 15/16	2 15/16
Rev. per Minute	33	33	33	33	33	33	33	33	33	33	33	33	33	33	33	33	33	33	33	33
Size Sprockets, In.	12¼	12¼	12¼	12¼	12¼	12¼	12¼	12¼	12¼	12¼	12¼	12¼	12¼	12¼	12¼	12¼	12¼	12¼	12¼	12¼
Diameter of Gear	17.84	17.84	17.84	17.84	17.84	29.83	29.83	29.83	29.83	29.83	29.83	29.83	29.83	29.83	29.83	29.83	29.83	29.83	35.82	35.82
Pitch of Gear	1	1	1	1	1	1¼	1¼	1¼	1¼	1¼	1¼	1¼	1¼	1¼	1¼	1¼	1¼	1¼	1½	1½
Face of Gear	2	2	2	2	2	3	3	3	3	3	3	3	3	3	3	3	3	3	4	4
Counter Shaft																				
Diameter, Inches	1 7/16	1 7/16	1 7/16	1 7/16	1 7/16	1 7/16	1 7/16	1 15/16	1 15/16	1 15/16	1 15/16	1 15/16	1 15/16	2 7/16	2 7/16	2 7/16	2 7/16	2 7/16	2 7/16	2 7/16
Rev. per Minute	115	115	115	115	115	165	165	165	165	165	165	165	165	165	165	165	165	165	165	165
Diam. of Pinion	5.12	5.12	5.12	5.12	5.12	6.01	6.01	6.01	6.01	6.01	6.01	6.01	6.01	6.01	6.01	6.01	6.01	6.01	7.22	7.22
Face of Pinion	2¾	2¾	2¾	2¾	2¾	3¼	3¼	3¼	3¼	3¼	3¼	3¼	3¼	3¼	3¼	3¼	3¼	3¼	4½	4½
Foot Shaft																				
Diameter, Inches	1 7/16	1 7/16	1 7/16	1 7/16	1 7/16	1 7/16	1 7/16	1 15/16	1 15/16	1 15/16	1 15/16	1 15/16	1 15/16	1 15/16	1 15/16	1 15/16	1 15/16	1 15/16	1 15/16	2 7/16
Size Sprockets, In.	12¼	12¼	12¼	12¼	12¼	12¼	12¼	12¼	12¼	12¼	12¼	12¼	12¼	12¼	12¼	12¼	12¼	12¼	12¼	12¼
Approx. Shpg. Wt.—lbs.																				
Terminals, Complete	565	595	625	645	700	680	710	935	965	1035	865	900	935	1115	1175	995	1030	1105	1275	1435
Apron Complete per Foot Centers	80	90	102	110	132	80	90	102	110	132	80	90	102	110	132	80	90	102	110	132

Material Weighing Approximately 100 Pounds per Cubic Foot

Length of Conveyor	0 to 25 ft. Centers				26 to 50 ft. Centers				51 to 75 ft. Centers				76 to 100 ft. Centers			
No. of Conveyor	2563	2583	2603	2623	2643	2663	2683	2703	2723	2743	2763	2783	2803	2823	2843	2863
Size of Material																
Average size of Material to be handled	4	7	9	14	4	7	9	14	4	7	9	14	4	7	9	14
Max. Size not to exceed 10% of Whole	8	14	18	24	8	14	18	24	8	14	18	24	8	14	18	24
Load in Lbs. per Foot on Conveyors	80	100	120	160	80	100	120	160	80	100	120	160	80	100	120	160
Capacity—Tons per Hour	240	298	360	480	240	298	360	480	240	298	360	480	240	298	360	480
Width of Apron	24	30	36	48	24	30	36	48	24	30	36	48	24	30	36	48
Horse Power at Center Shaft	2.3	2.8	3.2	4.0	4.7	5.5	6.4	8.1	7.0	8.3	9.6	12.1	9.3	11.0	12.8	16.2
Head Shaft																
Diameter, Inches	1 15/16	1 15/16	1 15/16	2 7/16	1 15/16	2 7/16	2 7/16	2 7/16	2 7/16	2 7/16	2 15/16	2 15/16	2 15/16	2 15/16	2 15/16	3 7/16
Rev. per Minute	33	33	33	33	33	33	33	33	33	33	33	33	33	33	33	33
Size Sprockets, Inches	12¼	12¼	12¼	12¼	12¼	12¼	12¼	12¼	12¼	12¼	12¼	12¼	12¼	12¼	12¼	12¼
Diameter of Gear	17.84	17.84	29.83	29.83	29.83	29.83	35.82	35.82	35.82	35.82	35.82	40.12	35.82	35.82	35.82	40.12
Pitch of Gear	1	1	1¼	1¼	1¼	1¼	1½	1½	1½	1½	1½	1½	1½	1½	1½	1½
Face of Gear	2	2	3	3	3	3	4	4	4	4	4	4	4	4	4	4
Counter Shaft																
Diameter, Inches	1 7/16	1 7/16	1 7/16	1 15/16	1 7/16	1 15/16	1 15/16	1 15/16	1 15/16	1 15/16	2 7/16	2 7/16	2 7/16	2 7/16	2 7/16	2 15/16
Rev. per Minute	115	115	165	165	165	165	165	165	165	165	165	185	165	165	165	185
Diameter of Pinion	5.12	5.12	6.01	6.01	6.01	6.01	7.22	7.22	7.22	7.22	7.22	7.22	7.22	7.22	7.22	7.22
Face of Pinion	2¾	2¾	3¼	3¼	3¼	3¼	4½	4½	4½	4½	4½	4½	4½	4½	4½	4½
Foot Shaft																
Diameter, Inches	1 7/16	1 7/16	1 7/16	1 15/16	1 7/16	1 15/16	1 15/16	1 15/16	1 15/16	1 15/16	2 7/16	2 7/16	2 7/16	2 7/16	2 7/16	2 7/16
Size Sprockets, Inches	12¼	12¼	12¼	12¼	12¼	12¼	12¼	12¼	12¼	12¼	12¼	12¼	12¼	12¼	12¼	12¼
Approx. Shipping Weight—Lbs.																
Terminals, Complete	595	625	760	1125	710	935	1125	1195	1020	1045	1345	1495	1215	1290	1350	1675
Apron Complete per Foot Centers	90	102	110	132	90	102	110	132	90	102	110	132	90	102	110	132

For Erection Dimensions of the above Conveyors, see opposite page.

General Dimensions of Jeffrey Standard Steel Apron Conveyors Using No. 126-C Malleable Roller Chain

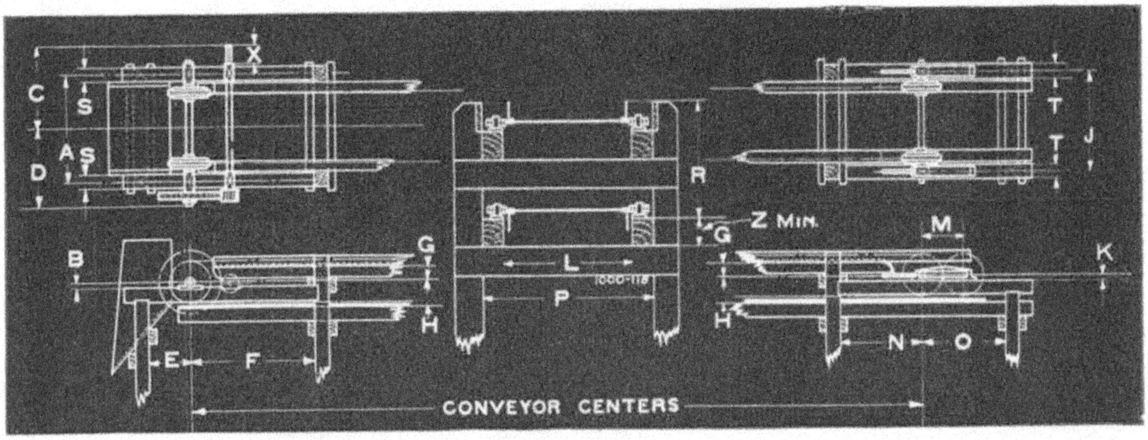

For Material Weighing 50 Pounds per Cubic Foot. Dimensions in Inches

Length of Conveyor	0 to 25 ft. Centers					26 to 50 ft. Centers					51 to 75 ft. Centers					76 to 100 ft. Centers				
No. of Conveyor	2197	2198	2221	2244	2267	2289	2290	2313	2336	2359	2381	2382	2405	2428	2451	2473	2474	2497	2520	2543
A	31 1/4	37 1/4	43 1/4	49 1/4	61 1/4	31 1/4	37 1/4	45 1/2	51 1/2	63 1/2	33 1/2	39 1/2	45 1/2	53 1/4	65 1/4	35 1/4	41 1/4	47 1/4	53 1/4	65 1/4
B	2 1/4	2 1/4	2 1/4	2 1/4	2 1/4	2 1/4	2 1/4	2 5/8	2 5/8	2 5/8	2 5/8	2 5/8	2 5/8	3 1/8	3 1/8	3 1/8	3 1/8	3 1/8	3 1/8	3 1/8
C	23 7/8	26 7/8	29 7/8	32 7/8	38 7/8	23 7/8	26 7/8	31 3/4	34 3/4	40 3/4	25 3/4	28 3/4	31 3/4	36 3/8	42 3/8	27 3/8	30 3/8	33 3/8	36 3/8	42 3/8
D	22 7/8	25 7/8	28 7/8	31 7/8	37 7/8	22 7/8	25 7/8	31 1/2	34 1/2	40 1/2	25 1/2	28 1/2	31 1/2	36 3/8	42 3/8	27 3/8	30 3/8	33 3/8	36 3/8	42 3/8
E	15	15	15	15	15	15	15	15	15	15	15	15	15	15	15	15	15	15	15	15
F	30	30	25	30	25	30	30	30	30	30	30	30	30	30	30	30	30	30	33	33
G	3 5/8	3 5/8	3 5/8	3 5/8	3 5/8	3 5/8	3 5/8	3 5/8	3 5/8	3 5/8	3 5/8	3 5/8	3 5/8	3 5/8	3 5/8	3 5/8	3 5/8	3 5/8	3 5/8	3 5/8
H	8 7/8	8 7/8	8 7/8	8 7/8	8 7/8	8 7/8	8 7/8	8 7/8	8 7/8	8 7/8	8 7/8	8 7/8	8 7/8	8 7/8	8 7/8	8 7/8	8 7/8	8 7/8	8 7/8	8 7/8
J	31 1/4	37 1/4	43 1/4	49 1/4	61 1/4	31 1/4	37 1/4	42 3/4	48 3/4	60 3/4	30 3/4	36 3/4	42 3/4	48 3/4	60 3/4	30 3/4	36 3/4	42 3/4	48 3/4	62 1/2
K	2 9/32	2 9/32	2 9/32	2 9/32	2 9/32	2 9/32	2 9/32	2 25/32	2 25/32	2 25/32	2 25/32	2 25/32	2 25/32	2 25/32	2 25/32	2 25/32	2 25/32	2 25/32	2 25/32	3 1/8
L	19 3/4	25 3/4	31 3/4	37 3/4	49 3/4	19 3/4	25 3/4	31 3/4	37 3/4	49 3/4	19 3/4	25 3/4	31 3/4	37 3/4	49 3/4	19 3/4	25 3/4	31 3/4	37 3/4	49 3/4
M	11 3/4	11 3/4	11 3/4	11 3/4	11 3/4	11 3/4	11 3/4	12	12	12	12	12	12	12	12	12	12	12	12	15
N	23	23	23	23	23	23	23	25	25	25	25	25	25	25	25	25	25	25	25	27
O	27	27	27	27	27	27	27	27	27	27	27	27	27	27	27	27	27	27	27	30
P	27 3/4	33 3/4	39 3/4	45 3/4	57 3/4	27 3/4	33 3/4	39 3/4	45 3/4	57 3/4	27 3/4	33 3/4	39 3/4	45 3/4	57 3/4	27 3/4	33 3/4	39 3/4	45 3/4	57 3/4
R	20 1/4	20 1/4	20 1/4	20 1/4	20 1/4	20 1/4	20 1/4	20 1/4	20 1/4	20 1/4	20 1/4	20 1/4	20 1/4	20 1/4	20 1/4	20 1/4	20 1/4	20 1/4	20 1/4	20 1/4
S	4	4	4	4	4	4	4	6	6	6	6	6	6	6	6	6	6	6	6	6
T	4	4	4	4	4	4	4	4	4	4	4	4	4	4	4	4	4	4	4	6
X	6	6	6	6	6	6	6	6	6	6	6	6	6	6	6	6	6	6	6	6
Z	6	6	6	6	6	6	6	6	6	6	6	6	6	6	6	6	6	6	6	6

For Material Weighing 100 Pounds per Cubic Foot. Dimensions in Inches

Length of Conveyor	0 to 25 ft. Centers				26 to 50 ft. Centers				51 to 75 ft. Centers				76 to 100 ft. Centers			
No. of Conveyor	2563	2583	2603	2623	2643	2663	2683	2703	2723	2743	2763	2783	2803	2823	2843	2863
A	37 1/4	43 1/4	49 1/4	63 1/2	37 1/4	45 1/2	51 1/2	63 1/2	39 1/2	45 1/2	53 1/4	65 1/4	41 1/4	47 1/4	53 1/4	67
B	2 1/4	2 1/4	2 1/4	2 5/8	2 1/4	2 5/8	2 5/8	2 5/8	2 5/8	2 5/8	3 1/8	3 1/8	3 1/8	3 1/8	3 1/8	3 1/2
C	26 7/8	29 7/8	32 7/8	40 3/4	26 7/8	31 3/4	34 3/4	40 3/4	28 3/4	31 3/4	36 3/8	42 3/8	30 3/8	33 3/8	36 3/8	44 5/8
D	25 7/8	28 7/8	31 7/8	40 1/2	25 7/8	31 1/2	34 1/2	40 1/2	28 1/2	31 1/2	36 3/8	42 3/8	30 3/8	33 3/8	36 3/8	44 1/4
E	15	15	15	15	15	15	15	15	15	15	15	15	15	15	15	15
F	30	25	30	30	30	30	30	33	30	33	33	36	33	33	33	36
G	3 5/8	3 5/8	3 5/8	3 5/8	3 5/8	3 5/8	3 5/8	3 5/8	3 5/8	3 5/8	3 5/8	3 5/8	3 5/8	3 5/8	3 5/8	3 5/8
H	8 7/8	8 7/8	8 7/8	8 7/8	8 7/8	8 7/8	8 7/8	8 7/8	8 7/8	8 7/8	8 7/8	8 7/8	8 7/8	8 7/8	8 7/8	8 7/8
J	37 1/4	43 1/4	49 1/4	60 3/4	37 1/4	42 3/4	48 3/4	60 3/4	36 3/4	43 3/4	50 1/2	62 1/2	38 1/2	44 1/2	50 1/2	62 1/2
K	2 9/32	2 9/32	2 9/32	2 25/32	2 9/32	2 25/32	2 25/32	2 25/32	2 25/32	2 25/32	3 1/8	3 1/8	3 1/8	3 1/8	3 1/8	3 1/8
L	25 3/4	31 3/4	37 3/4	49 3/4	25 3/4	31 3/4	37 3/4	49 3/4	25 3/4	31 3/4	37 3/4	49 3/4	25 3/4	31 3/4	37 3/4	49 3/4
M	11 3/4	11 3/4	11 3/4	12	11 3/4	12	12	12	12	12	15	15	15	15	15	15
N	23	23	23	25	23	25	25	25	25	25	27	27	27	27	27	27
O	27	27	27	27	27	27	27	27	27	27	30	30	30	30	30	30
P	33 3/4	39 3/4	45 3/4	57 3/4	33 3/4	39 3/4	45 3/4	57 3/4	33 3/4	39 3/4	45 3/4	57 3/4	33 3/4	39 3/4	45 3/4	57 3/4
R	20 1/4	20 1/4	20 1/4	20 1/4	20 1/4	20 1/4	20 1/4	20 1/4	20 1/4	20 1/4	20 1/4	20 1/4	20 1/4	20 1/4	20 1/4	20 1/4
S	4	4	4	6	4	6	6	6	6	6	6	6	6	6	6	8
T	4	4	4	4	4	4	4	4	4	4	6	6	6	6	6	6
X	6	6	6	6	6	6	6	6	6	6	6	6	6	6	6	7
Z	6	6	6	6	6	6	6	6	6	6	6	6	6	6	6	6

Specifications of Jeffrey Standard Steel Apron Conveyors Using No. 951 Steel Thimble Roller Chain

Material Weighing Approximately 50 Pounds per Cubic Foot

Length of Conveyor	0 to 25 ft. Centers				26 to 50 ft. Centers				51 to 75 ft. Centers				76 to 100 ft. Centers			
No. of Conveyor	2201	2224	2247	2270	2293	2316	2339	2362	2385	2408	2431	2454	2477	2500	2523	2546
Size of Material																
Average size of Material to be handled	4	7	9	14	4	7	9	14	4	7	9	14	4	7	9	14
Max. Size not to exceed 10% of Whole	8	14	18	24	8	14	18	24	8	14	18	24	8	14	18	24
Load in Lbs. per Foot on Conveyor	40	50	60	80	40	50	60	80	40	50	60	80	40	50	60	80
Capacity—Tons per Hour	120	149	180	240	120	149	180	240	120	149	180	240	120	149	180	240
Width of Apron	24	30	36	48	24	30	36	48	24	30	36	48	24	30	36	48
Horse Power at Center Shaft	1.6	1.8	2.1	2.6	3.2	3.7	4.2	5.1	4.8	5.5	6.3	7.8	6.4	7.3	8.3	10.3
Head Shaft																
Diameter, Inches	$1\frac{15}{16}$	$1\frac{15}{16}$	$1\frac{15}{16}$	$1\frac{15}{16}$	$1\frac{15}{16}$	$2\frac{7}{16}$	$2\frac{7}{16}$	$2\frac{7}{16}$	$2\frac{7}{16}$	$2\frac{7}{16}$	$2\frac{11}{16}$	$2\frac{11}{16}$	$2\frac{11}{16}$	$2\frac{11}{16}$	$2\frac{11}{16}$	$2\frac{11}{16}$
Rev. per Minute	33	33	30	33	33	33	33	33	33	33	33	33	33	33	33	33
Size Sprockets, Inches	$12\frac{3}{4}$	$12\frac{3}{4}$	$12\frac{3}{4}$	$12\frac{3}{4}$	$12\frac{3}{4}$	$12\frac{3}{4}$	$12\frac{3}{4}$	$12\frac{3}{4}$	$12\frac{3}{4}$	$12\frac{3}{4}$	$12\frac{3}{4}$	$12\frac{3}{4}$	$12\frac{3}{4}$	$12\frac{3}{4}$	$12\frac{3}{4}$	$12\frac{3}{4}$
Diameter of Gear	17.84	17.84	17.84	17.84	29.83	29.83	29.83	29.83	29.83	29.83	29.83	29.33	29.33	29.83	35.82	35.82
Pitch of Gear	1	1	1	1	$1\frac{1}{4}$	$1\frac{1}{4}$	$1\frac{1}{4}$	$1\frac{1}{4}$	$1\frac{1}{4}$	$1\frac{1}{4}$	$1\frac{1}{4}$	$1\frac{1}{4}$	$1\frac{1}{4}$	$1\frac{1}{4}$	$1\frac{1}{2}$	$1\frac{1}{2}$
Face of Gear	2	2	2	2	3	3	3	3	3	3	3	3	3	3	4	4
Counter Shaft																
Diameter, Inches	$1\frac{7}{16}$	$1\frac{7}{16}$	$1\frac{7}{16}$	$1\frac{7}{16}$	$1\frac{7}{16}$	$1\frac{11}{16}$	$1\frac{11}{16}$	$1\frac{11}{16}$	$1\frac{11}{16}$	$1\frac{11}{16}$	$2\frac{3}{16}$	$2\frac{3}{16}$	$2\frac{3}{16}$	$2\frac{3}{16}$	$2\frac{3}{16}$	$2\frac{3}{16}$
Rev. Per Minute	115½	115½	115½	115½	165	165	165	165	165	165	165	165	165	165	165	165
Diameter of Pinion	5.12	5.12	5.12	5.12	6.01	6.01	6.01	6.01	6.01	6.01	6.01	6.01	6.01	6.01	7.22	7.22
Face of Pinion	$2\frac{3}{4}$	$2\frac{3}{4}$	$2\frac{3}{4}$	$2\frac{3}{4}$	$3\frac{1}{4}$	$3\frac{1}{4}$	$3\frac{1}{4}$	$3\frac{1}{4}$	$3\frac{1}{4}$	$3\frac{1}{4}$	$3\frac{1}{4}$	$3\frac{1}{4}$	$3\frac{1}{4}$	$3\frac{1}{4}$	$4\frac{1}{4}$	$4\frac{1}{4}$
Foot Shaft																
Diameter, Inches	$1\frac{7}{16}$	$1\frac{7}{16}$	$1\frac{7}{16}$	$1\frac{7}{16}$	$1\frac{7}{16}$	$1\frac{11}{16}$	$1\frac{11}{16}$	$1\frac{11}{16}$	$1\frac{11}{16}$	$1\frac{11}{16}$	$1\frac{11}{16}$	$1\frac{11}{16}$	$1\frac{11}{16}$	$1\frac{11}{16}$	$1\frac{11}{16}$	$2\frac{3}{8}$
Size Sprockets, Inches	$12\frac{3}{4}$	$12\frac{3}{4}$	$12\frac{3}{4}$	$12\frac{3}{4}$	$12\frac{3}{4}$	$12\frac{3}{4}$	$12\frac{3}{4}$	$12\frac{3}{4}$	$12\frac{3}{4}$	$12\frac{3}{4}$	$12\frac{3}{4}$	$12\frac{3}{4}$	$12\frac{3}{4}$	$12\frac{3}{4}$	$12\frac{3}{4}$	$12\frac{3}{4}$
Approx. Shipping Weight—Lbs.																
Terminals, Complete	605	625	640	695	705	860	895	1015	875	915	1095	1175	1015	1060	1235	1400
Apron Complete per Foot Centers	102	114	122	144	102	114	122	144	102	114	122	144	102	114	122	144

Material Weighing Approximately 100 Pounds per Cubic Foot

Length of Conveyor	0 to 25 ft. Centers				26 to 50 ft. Centers				51 to 75 ft. Centers				76 to 100 ft. Centers			
No. of Conveyor	2566	2586	2606	2626	2646	2666	2686	2706	2726	2746	2766	2786	2806	2826	2846	2866
Size of Material																
Average size of Material to be handled	4	7	9	14	4	7	9	14	4	7	9	14	4	7	9	14
Max. Size not to exceed 10% of Whole	8	14	18	24	8	14	18	24	8	14	18	24	8	14	18	24
Load in Lbs. per Foot on Conveyor	80	100	120	160	80	100	120	160	80	100	120	160	80	100	120	160
Capacity—Tons per Hour	240	298	360	480	240	298	360	480	240	298	360	480	240	298	360	480
Width of Apron	24	30	36	48	24	30	36	48	24	30	36	48	24	30	36	48
Horse Power at Center Shaft	2.1	2.4	2.8	3.5	4.1	4.8	5.5	7.0	6.2	7.2	8.3	10.5	8.2	9.6	11.1	13.9
Head Shaft																
Diameter, Inches	$1\frac{15}{16}$	$1\frac{15}{16}$	$1\frac{15}{16}$	$2\frac{7}{16}$	$1\frac{15}{16}$	$2\frac{7}{16}$	$2\frac{7}{16}$	$2\frac{7}{16}$	$2\frac{7}{16}$	$2\frac{7}{16}$	$2\frac{11}{16}$	$2\frac{11}{16}$	$2\frac{11}{16}$	$2\frac{11}{16}$	$2\frac{11}{16}$	$3\frac{1}{16}$
Rev. per Minute	33	33	33	33	33	33	33	33	33	33	33	33	33	33	33	33
Size Sprockets, Inches	$12\frac{3}{4}$	$12\frac{3}{4}$	$12\frac{3}{4}$	$12\frac{3}{4}$	$12\frac{3}{4}$	$12\frac{3}{4}$	$12\frac{3}{4}$	$12\frac{3}{4}$	$12\frac{3}{4}$	$12\frac{3}{4}$	$12\frac{3}{4}$	$12\frac{3}{4}$	$12\frac{3}{4}$	$12\frac{3}{4}$	$12\frac{3}{4}$	$12\frac{3}{4}$
Diameter of Gear	17.84	17.84	29.83	29.83	29.83	29.83	35.82	35.82	35.82	35.82	35.82	35.82	35.82	35.82	35.82	40.12
Pitch of Gear	1	1	$1\frac{1}{4}$	$1\frac{1}{4}$	$1\frac{1}{4}$	$1\frac{1}{4}$	$1\frac{1}{2}$	$1\frac{1}{2}$	$1\frac{1}{2}$	$1\frac{1}{2}$	$1\frac{1}{2}$	$1\frac{1}{2}$	$1\frac{1}{2}$	$1\frac{1}{2}$	$1\frac{1}{2}$	$1\frac{1}{2}$
Face of Gear	2	2	3	3	3	3	4	4	4	4	4	4	4	4	4	$4\frac{1}{2}$
Counter Shaft																
Diameter, Inches	$1\frac{7}{16}$	$1\frac{7}{16}$	$1\frac{7}{16}$	$1\frac{11}{16}$	$1\frac{7}{16}$	$1\frac{11}{16}$	$1\frac{11}{16}$	$1\frac{11}{16}$	$1\frac{11}{16}$	$1\frac{11}{16}$	$2\frac{3}{16}$	$2\frac{3}{16}$	$2\frac{3}{16}$	$2\frac{3}{16}$	$2\frac{3}{16}$	$2\frac{11}{16}$
Rev. per Minute	115	115	165	165	165	165	165	165	165	165	165	185	165	165	165	165
Diameter of Pinion	5.12	5.12	6.01	6.01	6.01	6.01	7.22	7.22	7.22	7.22	7.22	7.22	7.22	7.22	7.22	7.22
Face of Pinion	$2\frac{3}{4}$	$2\frac{3}{4}$	$3\frac{1}{4}$	$3\frac{1}{4}$	$3\frac{1}{4}$	$3\frac{1}{4}$	$4\frac{1}{4}$	$4\frac{1}{4}$	$4\frac{1}{4}$	$4\frac{1}{4}$	$4\frac{1}{4}$	$4\frac{1}{4}$	$4\frac{1}{4}$	$4\frac{1}{4}$	$4\frac{1}{4}$	$4\frac{1}{4}$
Foot Shaft																
Diameter, Inches	$1\frac{7}{16}$	$1\frac{7}{16}$	$1\frac{7}{16}$	$1\frac{11}{16}$	$1\frac{7}{16}$	$1\frac{11}{16}$	$1\frac{11}{16}$	$1\frac{11}{16}$	$1\frac{11}{16}$	$1\frac{11}{16}$	$2\frac{3}{16}$	$2\frac{3}{16}$	$2\frac{3}{16}$	$2\frac{3}{16}$	$2\frac{3}{16}$	$2\frac{3}{8}$
Size Sprockets, Inches	$12\frac{3}{4}$	$12\frac{3}{4}$	$12\frac{3}{4}$	$12\frac{3}{4}$	$12\frac{3}{4}$	$12\frac{3}{4}$	$12\frac{3}{4}$	$12\frac{3}{4}$	$12\frac{3}{4}$	$12\frac{3}{4}$	$12\frac{3}{4}$	$12\frac{3}{4}$	$12\frac{3}{4}$	$12\frac{3}{4}$	$12\frac{3}{4}$	$12\frac{3}{4}$
Approx. Shipping Weight—Lbs.																
Terminals, Complete	605	625	775	1050	705	860	1140	1210	1075	1110	1360	1505	1245	1320	1360	1690
Apron Complete per Foot Centers	102	114	122	144	102	114	122	144	102	114	122	144	102	114	122	144

For Erection Dimensions of the above Conveyors, see Opposite page.

JEFFREY — *Steel Apron Conveyors*

General Dimensions of Jeffrey Standard Steel Apron Conveyors Using No. 951 Steel Thimble Roller Chain

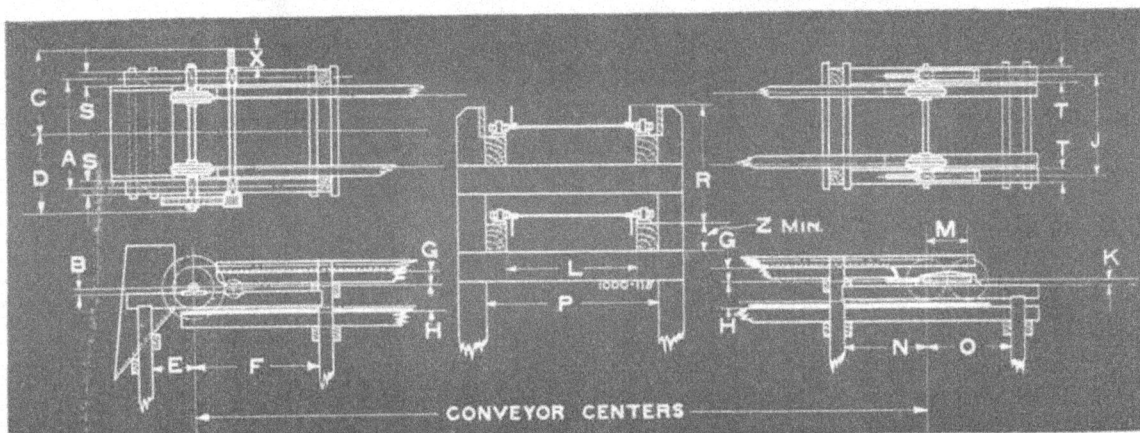

For Material Weighing 50 Pounds per Cubic Foot. Dimensions in Inches

Length of Conveyor	0 to 25 ft. Centers				26 to 50 ft. Centers				51 to 75 ft. Centers				76 to 100 ft. Centers			
No. of Conveyor	2201	2224	2247	2270	2293	2316	2339	2362	2385	2408	2431	2454	2477	2500	2523	2546
A	37¼	43¼	49¼	61¼	37¼	45½	51½	63½	39½	45½	53¼	65¼	41¼	47¼	53¼	65¼
B	2¼	2¼	2¼	2¼	2¼	2⅝	2⅝	2⅝	2⅝	2⅝	3⅛	3⅛	3⅛	3⅛	3⅛	3⅛
C	26⅞	29⅞	32⅞	38⅞	26⅞	31¾	34¾	40¾	28¾	31¾	36⅜	42⅜	30⅜	33⅜	36⅜	42⅜
D	25⅞	28⅞	31⅞	37⅞	25⅞	31½	34½	40½	28½	31½	36⅜	42⅜	30⅜	33⅜	36⅜	42⅜
E	15	15	15	15	15	15	15	15	15	15	15	15	15	15	15	15
F	30	25	30	25	30	30	30	30	30	30	30	30	30	30	33	33
G	3⅝	3⅝	3⅝	3⅝	3⅝	3⅝	3⅝	3⅝	3⅝	3⅝	3⅝	3⅝	3⅝	3⅝	3⅝	3⅝
H	8⅞	8⅞	8⅞	8⅞	8⅞	8⅞	8⅞	8⅞	8⅞	8⅞	8⅞	8⅞	8⅞	8⅞	8⅞	8⅞
J	37¼	43¼	49¼	61¼	37¼	42¾	48¾	60¾	36¾	42¾	48¾	60¾	36¾	42¾	48¾	62½
K	2 9/32	2 9/32	2 9/32	2 9/32	2 9/32	2 25/32	2 25/32	2 25/32	2 25/32	2 25/32	2 25/32	2 25/32	2 25/32	2 25/32	2 25/32	3⅛
L	25¾	31¾	37¾	49¾	25¾	31¾	37¾	49¾	25¾	31¾	37¾	49¾	25¾	31¾	37¾	49¾
M	11¾	11¾	11¾	11¾	11¾	12	12	12	12	12	12	12	12	12	12	15
N	23	23	23	23	23	25	25	25	25	25	25	25	25	25	25	27
O	27	27	27	27	27	27	27	27	27	27	27	27	27	27	27	30
P	33¾	39¾	45¾	57¾	33¾	39¾	45¾	57¾	33¾	39¾	45¾	57¾	33¾	39¾	45¾	57¾
R	20¼	20¼	20¼	20¾	20¼	20¼	20¾	20¼	20¼	20¼	20¼	20¼	20¼	20¼	20¼	20¼
S	4	4	4	4	4	6	6	6	6	6	6	6	6	6	6	6
T	4	4	4	4	4	4	4	4	4	4	4	4	4	4	4	6
X	6	6	6	6	6	6	6	6	6	6	6	6	6	6	6	6
Z	6	6	6	6	6	6	6	6	6	6	6	6	6	6	6	6

For Material Weighing 100 Pounds per Cubic Foot. Dimensions in Inches

Length of Conveyor	0 to 25 ft. Centers				26 to 50 ft. Centers				51 to 75 ft. Centers				76 to 100 ft. Centers			
No. of Conveyor	2566	2586	2606	2626	2646	2666	2686	2706	2726	2746	2766	2786	2806	2826	2846	2866
A	37¼	43¼	49¼	63½	37¼	45½	51½	63½	39½	45½	53¼	65¼	41¼	47¼	53¼	67
B	2¼	2¼	2¼	2⅝	2¼	2⅝	2⅝	2⅝	2⅝	2⅝	3⅛	3⅛	3⅛	3⅛	3⅛	3½
C	26⅞	29⅞	32⅞	40¾	26⅞	31¾	34¾	40¾	28¾	31¾	36⅜	42⅜	30⅜	33⅜	36⅜	44⅝
D	25⅞	28⅞	31⅞	40½	25⅞	31½	34½	40½	28½	31½	36⅜	42⅜	30⅜	33⅜	36⅜	44⅜
E	15	15	15	15	15	15	15	15	15	15	15	15	15	15	15	15
F	30	25	30	30	30	30	30	33	30	33	33	36	33	33	33	36
G	3⅝	3⅝	3⅝	3⅝	3⅝	3⅝	3⅝	3⅝	3⅝	3⅝	3⅝	3⅝	3⅝	3⅝	3⅝	3⅝
H	8⅞	8⅞	8⅞	8⅞	8⅞	8⅞	8⅞	8⅞	8⅞	8⅞	8⅞	8⅞	8⅞	8⅞	8⅞	8⅞
J	37¼	43¼	49¼	60¾	37¼	42¾	48¾	60¾	36¾	42¾	50½	62½	38½	44½	50½	62½
K	2 9/32	2 9/32	2 9/32	2 25/32	2 9/32	2 25/32	2 25/32	2 25/32	2 25/32	2 25/32	3⅛	3⅛	3⅛	3⅛	3⅛	3⅛
L	25¾	31¾	37¾	49¾	25¾	31¾	37¾	49¾	25¾	31¾	37¾	49¾	25¾	31¾	37¾	49¾
M	11¾	11¾	11¾	12	11¾	12	12	12	12	12	15	15	15	15	15	15
N	23	23	23	25	23	25	25	25	25	25	27	27	27	27	27	27
O	27	27	27	27	27	27	27	27	27	27	30	30	30	30	30	30
P	33¾	39¾	45¾	57¾	33¾	39¾	45¾	57¾	33¾	39¾	45¾	57¾	33¾	39¾	45¾	57¾
R	20¼	20¼	20¼	20¼	20¼	20¼	20¼	20¼	20¼	20¼	20¼	20¼	20¼	20¼	20¼	20¼
S	4	4	4	6	4	6	6	6	6	6	6	6	6	6	6	8
T	4	4	4	4	4	4	4	4	4	4	6	6	6	6	6	6
X	6	6	6	6	6	6	6	6	6	6	6	6	6	6	6	7
Z	6	6	6	6	6	6	6	6	6	6	6	6	6	6	6	6

Specifications of Jeffrey Standard Steel Apron Conveyors Using No. 809 Steel Thimble Roller Chain

Material Weighing Approximately 50 Pounds per Cubic Foot

Length of Conveyor	0 to 25 ft. Centers				26 to 50 ft. Centers				51 to 75 ft. Centers				76 to 100 ft. Centers			
No. of Conveyor	2204	2227	2250	2273	2296	2319	2342	2365	2388	2411	2434	2457	2480	2503	2526	2549
Size of Material																
Average size of Material to be handled	7	9	14	14	7	9	14	14	7	9	14	14	7	9	14	14
Max. Size not to exceed 10% of Whole	14	18	24	24	14	18	24	24	14	18	24	24	14	18	24	24
Load in Lbs. per Foot on Conveyor	50	60	80	100	50	60	80	100	50	60	80	100	50	60	80	100
Capacity—Tons per Hour	149	180	240	300	149	180	240	300	149	180	240	300	149	180	240	300
Width of Apron	30	36	48	60	30	36	48	60	30	36	48	60	30	36	48	60
Horse Power at Center Shaft	2.3	2.5	3.1	3.5	4.7	5.1	6.1	7.1	7.0	7.7	9.3	10.6	9.3	10.2	12.2	14.2
Head Shaft																
Diameter, Inches	1 11/16	1 11/16	2 7/16	2 7/16	2 7/16	2 7/16	2 11/16	2 11/16	2 11/16	2 11/16	2 11/16	2 11/16	2 11/16	3 7/16	3 7/16	3 7/16
Rev. per Minute	22	22	22	22	22	22	22	22	22	22	22	22	22	22	22	22
Size Sprockets, Inches	18	18	18	18	18	18	18	18	18	18	18	18	18	18	18	18
Diameter of Gear	29.83	29.83	29.83	29.83	29.83	29.83	29.83	29.83	29.83	35.82	35.82	35.82	35.82	35.82	36.78	36.78
Pitch of Gear	1¼	1¼	1¼	1¼	1¼	1¼	1½	1¼	1¼	1½	1½	1½	1½	1½	1¾	1¾
Face of Gear	3	3	3	3	3	3	3	3	3	4	4	4	4	4	5½	5½
Counter Shaft																
Diameter, Inches	1 7/16	1 7/16	1 11/16	1 11/16	1 11/16	1 11/16	2 7/16	2 7/16	2 7/16	2 7/16	2 7/16	2 7/16	2 7/16	2 11/16	2 11/16	2 11/16
Rev. per Minute	110	110	110	110	110	110	110	110	110	110	110	110	110	110	103	103
Diameter of Pinion	6.01	6.01	6.01	6.01	6.01	6.01	6.01	6.01	6.01	7.22	7.22	7.22	7.22	7.22	7.86	7.86
Face of Pinion	3¼	3¼	3¼	3¼	3¼	3¼	3¼	3¼	3¼	4½	4½	4½	4½	4½	6	6
Foot Shaft																
Diameter, Inches	1 7/16	1 7/16	1 11/16	1 11/16	1 11/16	1 11/16	1 11/16	1 11/16	1 11/16	1 11/16	2 7/16	2 7/16	2 7/16	2 7/16	2 7/16	2 7/16
Size Sprockets, Inches	18	18	18	18	18	18	18	18	18	18	18	18	18	18	18	18
Approx. Shipping Weight—Lbs.																
Terminals, Complete	1225	1265	1570	1670	1415	1465	1740	1850	1565	1780	1980	2095	1800	2015	2355	2485
Apron Complete per Foot Centers	190	204	234	262	190	204	234	262	190	204	234	262	190	204	234	262

Material Weighing Approximately 100 Pounds per Cubic Foot

Length of Conveyor	0 to 25 ft. Centers				26 to 50 ft. Centers				51 to 75 ft. Centers				76 to 100 ft. Centers			
No. of Conveyor	2569	2589	2609	2629	2649	2669	2689	2709	2729	2749	2769	2789	2809	2829	2849	2869
Size of Material																
Average size of Material to be handled	7	9	14	14	7	9	14	14	7	9	14	14	7	9	14	14
Max. Size not to exceed 10% of Whole	14	18	24	24	14	18	24	24	14	18	24	24	14	18	24	24
Load in Lbs. per Foot on Conveyor	100	120	160	200	100	120	160	200	100	120	160	200	100	120	160	200
Capacity—Tons per Hour	298	360	480	600	298	360	480	600	298	360	480	600	298	360	480	600
Width of Apron	30	36	48	60	30	36	48	60	30	36	48	60	30	36	48	60
Horse Power at Center Shaft	2.9	3.2	4.0	4.7	5.8	6.5	8.0	9.4	8.7	9.7	11.9	14.1	11.5	13.0	15.9	18.8
Head Shaft																
Diameter, Inches	2 7/16	2 7/16	2 7/16	2 11/16	2 7/16	2 11/16	2 11/16	2 11/16	2 11/16	2 11/16	3 7/16	3 7/16	3 7/16	3 7/16	3 7/16	3 11/16
Rev. per Minute	22	22	22	22	22	22	22	22	22	22	22	22	22	22	22	22
Size Sprockets, Inches	18	18	18	18	18	18	18	18	18	18	18	18	18	18	18	18
Diameter of Gear	29.83	29.83	29.83	29.83	35.82	35.82	35.82	40.12	35.82	35.82	35.82	36.78	35.82	40.12	36.78	36.78
Pitch of Gear	1¼	1¼	1¼	1¼	1½	1½	1½	1½	1½	1½	1½	1¾	1½	1½	1¾	1¾
Face of Gear	3	3	3	3	4	4	4	4	4	4	4	5½	4	4	5½	5½
Counter Shaft																
Diameter, Inches	1 11/16	1 11/16	1 11/16	2 7/16	1 11/16	2 7/16	2 7/16	2 7/16	2 7/16	2 7/16	2 11/16	2 11/16	2 11/16	2 11/16	2 11/16	2 11/16
Rev. per Minute	110	110	110	110	110	110	110	123	110	110	110	103	110	123	103	103
Diameter of Pinion	6.01	6.01	6.01	6.01	7.22	7.22	7.22	7.22	7.22	7.22	7.22	7.86	7.22	7.22	7.86	7.86
Face of Pinion	3¼	3¼	3¼	3¼	4½	4½	4½	4½	4½	4½	4½	6	4½	4½	6	6
Foot Shaft																
Diameter, Inches	1 11/16	1 11/16	1 11/16	2 7/16	1 11/16	2 7/16	2 7/16	2 7/16	2 7/16	2 7/16	2 7/16	2 7/16	2 7/16	2 7/16	2 7/16	2 11/16
Size Sprockets, Inches	18	18	18	18	18	18	18	18	18	18	18	18	18	18	18	18
Approx. Shipping Weight—Lbs.																
Terminals, Complete	1415	1465	1510	1935	1635	1855	1980	2155	1800	1855	2150	2485	1955	2075	2355	2920
Apron Complete per Foot Centers	190	204	234	262	190	204	234	262	190	204	234	262	190	204	234	262

For Erection Dimensions of the above Conveyors, see Opposite page.

General Dimensions of Jeffrey Standard Steel Apron Conveyors Using No. 809 Steel Thimble Roller Chain

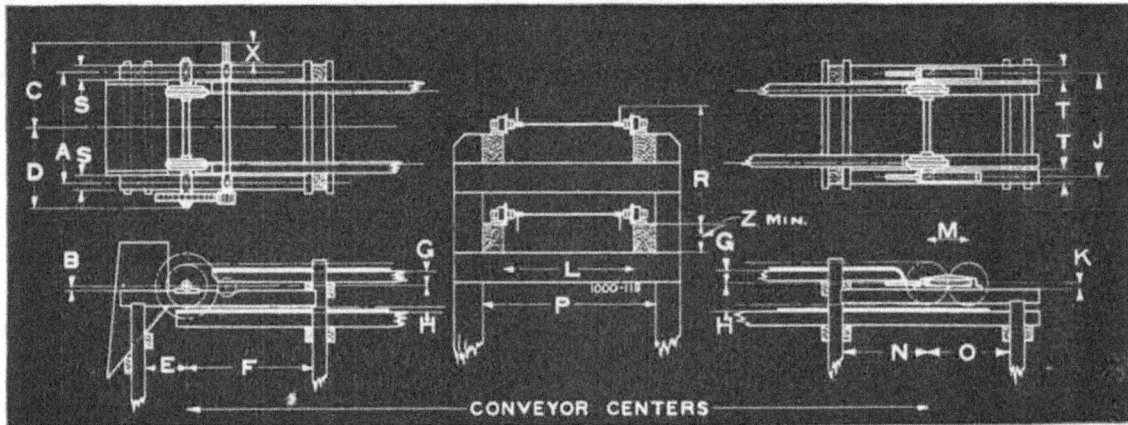

For Material Weighing 50 Pounds per Cubic Foot. Dimensions in Inches

Length of Conveyor	0 to 25 ft. Centers				26 to 50 ft. Centers				51 to 75 ft. Centers				76 to 100 ft. Centers			
No. of Conveyor	2204	2227	2250	2273	2296	2319	2342	2365	2388	2411	2434	2457	2480	2503	2526	2549
A	43¾	49¾	64	76	46	52	65¾	77¾	47¾	53¾	65¾	77¾	47¾	55½	67½	79½
B	2¼	2¼	2⅝	2⅝	2⅝	2⅝	3⅛	3⅛	3⅛	3⅛	3⅛	3⅛	3⅛	3½	3½	3½
C	30⅛	33⅛	41	47	32	35	42⅝	48⅝	33⅝	36⅝	42⅝	48⅝	33⅝	38⅞	44⅞	50⅞
D	29⅛	32⅛	40¾	46¾	31¾	34¾	42⅝	48⅝	33⅝	36⅝	42⅝	48⅝	33⅝	38½	44½	50½
E	18	18	18	18	18	18	18	18	18	18	18	18	18	18	18	18
F	30	30	30	30	30	30	30	30	30	33	33	33	33	33	35	35
G	5¾	5¾	5¾	5¾	5¾	5¾	5¾	5¾	5¾	5¾	5¾	5¾	5¾	5¾	5¾	5¾
H	12	12	12	12	12	12	12	12	12	12	12	12	12	12	12	12
J	43¾	49¾	61¼	73¼	43¼	49¼	61¼	73¼	43¼	49¼	63	75	45	51	63	75
K	2 9/32	2 9/32	2 29/32	2 29/32	2 29/32	2 29/32	2 29/32	2 29/32	2 29/32	2 29/32	3⅛	3⅛	3⅛	3⅛	3⅛	3⅜
L	32⅝	38⅝	50⅝	62⅝	32⅝	38⅝	50⅝	62⅝	32⅝	38⅝	50⅝	62⅝	32⅝	38⅝	50⅝	62⅝
M	11¾	11¾	12	12	12	12	12	12	12	12	15	15	15	15	15	15
N	23	23	25	25	25	25	25	25	25	25	27	27	27	27	27	27
O	30	30	30	30	30	30	30	30	30	30	33	33	33	33	33	33
P	40⅝	46⅝	58⅝	70⅝	40⅝	46⅝	58⅝	70⅝	40⅝	46⅝	58⅝	70⅝	40⅝	46⅝	58⅝	70⅝
R	25¾	25¾	25¾	25¾	25¾	25¾	25¾	25¾	25¾	25¾	25¾	25¾	25¾	25¾	25¾	25¾
S	4	4	6	6	6	6	6	6	6	6	6	6	6	8	8	8
T	4	4	4	4	4	4	4	4	4	4	6	6	6	6	6	6
X	6	6	6	6	6	6	6	6	6	6	6	6	6	7	7	7
Z	6	6	6	6	6	6	6	6	6	6	6	6	6	6	6	6

For Material Weighing 100 Pounds per Cubic Foot. Dimensions in Inches

Length of Conveyor	0 to 25 ft. Centers				26 to 50 ft. Centers				51 to 75 ft. Centers				76 to 100 ft. Centers			
No. of Conveyor	2569	2589	2609	2629	2649	2669	2689	2709	2729	2749	2769	2789	2809	2829	2849	2869
A	46	52	64	77¾	46	53¾	65¾	77¾	47¾	53¾	67½	79½	49½	55½	67½	81¾
B	2⅝	2⅝	2⅝	3⅛	2⅝	3⅛	3⅛	3⅛	3⅛	3⅛	3½	3½	3½	3½	3½	4⅛
C	32	35	41	48⅝	32	36⅝	42⅝	48⅝	33⅝	36⅝	44⅞	50⅞	35⅞	38⅞	44⅞	53⅜
D	31¾	34¾	40¾	48⅝	31¾	36⅝	42⅝	48⅝	33⅝	36⅝	44½	50½	35½	38½	44½	53⅜
E	18	18	18	18	18	18	18	18	18	18	18	18	18	18	18	18
F	30	30	30	30	30	36	30	36	33	33	33	35	33	33	35	35
G	5¾	5¾	5¾	5¾	5¾	5¾	5¾	5¾	5¾	5¾	5¾	5¾	5¾	5¾	5¾	5¾
H	12	12	12	12	12	12	12	12	12	12	12	12	12	12	12	12
J	43¼	49¾	61¼	75	43¼	51	63	75	45	51	63	75	45	51¼	63	76¾
K	2 29/32	2 29/32	2 29/32	3⅛	2 29/32	3⅛	3⅛	3⅛	3⅛	3⅛	3⅛	3⅛	3⅛	3⅛	3⅛	4
L	32⅝	38⅝	50⅝	62⅝	32⅝	38⅝	50⅝	62⅝	32⅝	38⅝	50⅝	62⅝	32⅝	38⅝	50⅝	62⅝
M	12	12	12	15	12	15	15	15	15	15	15	15	15	15	15	10½
N	25	25	25	27	25	27	27	27	27	27	27	27	27	27	27	25
O	30	30	30	33	30	33	33	33	33	33	33	33	33	33	33	29
P	40⅝	46⅝	58⅝	70⅝	40⅝	46⅝	58⅝	70⅝	40⅝	46⅝	58⅝	70⅝	40⅝	46⅝	58⅝	70⅝
R	25¾	25¾	25¾	25¾	25¾	25¾	25¾	25¾	25¾	25¾	25¾	25¾	25¾	25¾	25¾	25¾
S	6	6	6	6	6	6	6	6	6	6	8	8	8	6	8	8
T	4	4	4	6	4	6	6	6	6	6	6	6	6	6	6	6
X	6	6	6	6	6	6	6	6	6	6	7	7	7	7	7	8
Z	6	6	6	6	6	6	6	6	6	6	6	6	6	6	6	6

Specifications of Jeffrey Standard Steel Apron Conveyors Using No. 276 Steel Thimble Roller Chain

Material Weighing Approximately 50 Pounds per Cubic Foot

Length of Conveyor	0 to 25 ft. Centers				26 to 50 ft. Centers				51 to 75 ft. Centers				76 to 100 ft. Centers			
No. of Conveyor	**2206**	**2229**	**2252**	**2275**	**2298**	**2321**	**2344**	**2367**	**2390**	**2413**	**2436**	**2459**	**2482**	**2505**	**2528**	**2551**
Size of Material																
Average size of Material to be handled	7	9	14	14	7	9	14	14	7	9	14	14	7	9	14	14
Max. Size not to exceed 10% of Whole	14	18	24	24	14	18	24	24	14	18	24	24	14	18	24	24
Load in Lbs. per Foot on Conveyor	50	60	80	100	50	60	80	100	50	60	80	100	50	60	80	100
Capacity—Tons per Hour	149	180	240	300	149	180	240	300	149	180	240	300	149	180	240	300
Width of Apron	30	36	48	60	30	36	48	60	30	36	48	60	30	36	48	60
Horse Power at Center Shaft	1.9	2.1	2.6	3.0	3.7	4.2	5.1	6.1	6.0	6.3	7.7	9.1	7.5	8.4	10.2	12
Head Shaft																
Diameter, Inches	1 15/16	1 15/16	2 7/16	2 7/16	2 7/16	2 7/16	2 15/16	2 15/16	2 15/16	2 15/16	2 15/16	2 15/16	2 15/16	3 7/16	3 7/16	3 7/16
Rev. per Minute	16 2/3	16 2/3	16 2/3	16 2/3	16 2/3	16 2/3	16 2/3	16 2/3	16 2/3	16 2/3	16 2/3	16 2/3	16 2/3	16 2/3	16 2/3	16 2/3
Size Sprockets, Inches	24	24	24	24	24	24	24	24	24	24	24	24	24	24	24	24
Diameter of Gear	29.83	29.83	29.83	29.83	29.83	29.83	29.83	29.83	29.83	35.82	35.82	35.82	35.82	35.82	36.78	36.78
Pitch of Gear	1 1/4	1 1/4	1 1/4	1 1/4	1 1/4	1 1/4	1 1/4	1 1/4	1 1/4	1 1/2	1 1/2	1 1/2	1 1/2	1 1/2	1 3/4	1 3/4
Face of Gear	3	3	3	3	3	3	3	3	3	4	4	4	4	4	5 1/2	5 1/2
Counter Shaft																
Diameter, Inches	1 7/16	1 7/16	1 15/16	1 15/16	1 15/16	1 15/16	2 7/16	2 7/16	2 7/16	2 7/16	2 7/16	2 7/16	2 7/16	2 11/16	2 11/16	2 11/16
Rev. per Minute	83	83	83	83	83	83	83	83	83	83	83	83	83	83	78	78
Diameter of Pinion	6.01	6.01	6.01	6.01	6.01	6.01	6.01	6.01	6.01	7.22	7.22	7.22	7.22	7.22	7.86	7.86
Face of Pinion	3 1/4	3 1/4	3 1/4	3 1/4	3 1/4	3 1/4	3 1/4	3 1/4	3 1/4	4 1/2	4 1/2	4 1/2	4 1/2	4 1/2	6	6
Foot Shaft																
Diameter, Inches	1 7/16	1 7/16	1 15/16	1 15/16	1 15/16	1 15/16	1 15/16	1 15/16	1 15/16	1 15/16	2 7/16	2 7/16	2 7/16	2 7/16	2 7/16	2 7/16
Size Sprockets, Inches	24	24	24	24	24	24	24	24	24	24	24	24	24	24	24	24
Approx. Shipping Weight—Lbs.																
Terminals, Complete	1535	1590	1965	2020	1730	1790	2065	2190	1870	2095	2315	2435	2105	2320	2660	2815
Apron Complete per Foot Centers	140	154	180	206	140	154	180	206	140	154	180	206	140	154	180	206

Material Weighing Approximately 100 Pounds per Cubic Foot

Length of Conveyor	0 to 25 ft. Centers				26 to 50 ft. Centers				51 to 75 ft. Centers				76 to 100 ft. Centers			
No. of Conveyor	**2571**	**2591**	**2611**	**2631**	**2651**	**2671**	**2691**	**2711**	**2731**	**2751**	**2771**	**2791**	**2811**	**2831**	**2851**	**2871**
Size of Material																
Average size of Material to be handled	7	9	14	14	7	9	14	14	7	9	14	14	7	9	14	14
Max. Size not to exceed 10% of Whole	14	18	24	24	14	18	24	24	14	18	24	24	14	18	24	24
Load in Lbs. per Foot on Conveyor	100	120	160	200	100	120	160	200	100	120	160	200	100	120	160	200
Capacity—Tons per Hour	298	360	480	600	298	360	480	600	298	360	480	600	298	360	480	600
Width of Apron	30	36	48	60	30	36	48	60	30	36	48	60	30	36	48	60
Horse Power at Center Shaft	2.4	2.8	3.7	4.2	4.9	5.6	7.0	8.4	7.3	8.4	10.5	12.5	9.7	11.1	13.9	16.7
Head Shaft																
Diameter, Inches	2 7/16	2 7/16	2 7/16	2 15/16	2 7/16	2 15/16	2 15/16	2 15/16	2 15/16	2 15/16	3 7/16	3 7/16	3 7/16	3 7/16	3 7/16	3 15/16
Rev. per Minute	16 2/3	16 2/3	16 2/3	16 2/3	16 2/3	16 2/3	16 2/3	16 2/3	16 2/3	16 2/3	16 2/3	16 2/3	16 2/3	16 2/3	16 2/3	16 2/3
Size Sprockets, Inches	24	24	24	24	24	24	24	24	24	24	24	24	24	24	24	24
Diameter of Gear	29.83	29.83	29.83	29.83	35.82	35.82	35.82	40.12	35.82	35.82	35.82	36.78	35.82	40.12	36.78	36.78
Pitch of Gear	1 1/4	1 1/4	1 1/4	1 1/4	1 1/2	1 1/2	1 1/2	1 1/2	1 1/2	1 1/2	1 1/2	1 3/4	1 1/2	1 1/2	1 3/4	1 3/4
Face of Gear	3	3	3	3	4	4	4	4	4	4	4	5 1/2	4	4	5 1/2	5 1/2
Counter Shaft																
Diameter, Inches	1 15/16	1 15/16	1 15/16	2 7/16	1 15/16	2 7/16	2 7/16	2 7/16	2 7/16	2 7/16	2 11/16	2 11/16	2 11/16	2 11/16	2 11/16	2 11/16
Rev per Minute	83	83	83	83	83	83	83	93	83	83	83	78	83	93	78	78
Diameter of Pinion	6.01	6.01	6.01	6.01	7.22	7.22	7.22	7.22	7.22	7.22	7.22	7.86	7.22	7.22	7.86	7.86
Face of Pinion	3 1/4	3 1/4	3 1/4	3 1/4	4 1/2	4 1/2	4 1/2	4 1/2	4 1/2	4 1/2	4 1/2	6	4 1/2	4 1/2	6	6
Foot Shaft																
Diameter, Inches	1 15/16	1 15/16	1 15/16	2 7/16	1 15/16	2 7/16	2 7/16	2 7/16	2 7/16	2 7/16	2 7/16	2 7/16	2 7/16	2 7/16	2 7/16	2 11/16
Size Sprockets, Inches	24	24	24	24	24	24	24	24	24	24	24	24	24	24	24	24
Approx. Shipping Weight—Lbs.																
Terminals, Complete	1730	1790	1965	2280	1890	2180	2315	2500	2105	2180	2465	2815	2245	2380	2670	3230
Apron Complete per Foot Centers	140	154	180	206	140	154	180	206	140	154	180	206	140	154	180	206

For Erection Dimensions of the above Conveyors, see Opposite page.

General Dimensions of Jeffrey Standard Steel Apron Conveyors Using No. 276 Steel Thimble Roller Chain

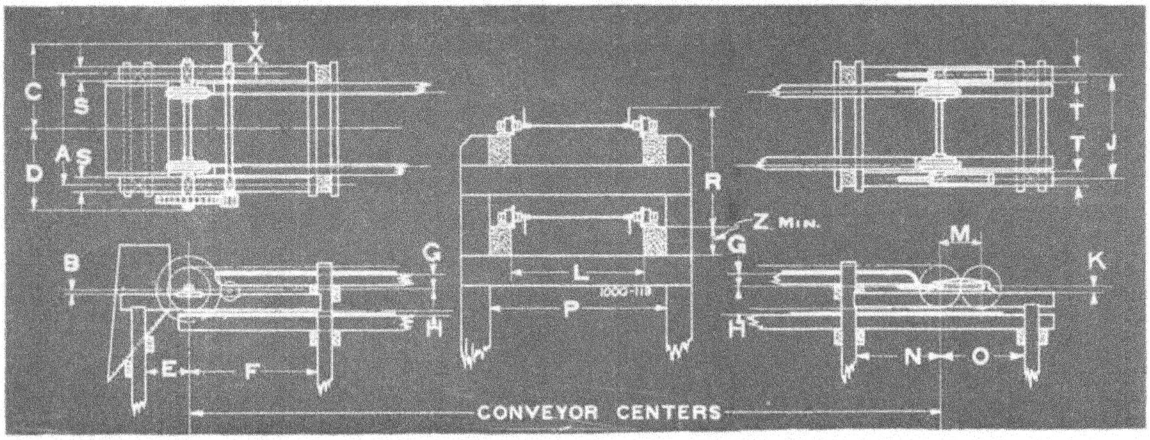

For Material Weighing 50 Pounds per Cubic Foot. Dimensions in Inches

Length of Conveyor	0 to 25 ft. Centers				26 to 50 ft. Centers				51 to 75 ft. Centers				76 to 100 ft. Centers			
No. of Conveyor	2206	2229	2252	2275	2298	2321	2344	2367	2390	2413	2436	2459	2482	2505	2528	2551
A	44⅛	50⅛	64⅜	76⅜	46⅜	52⅜	66⅛	78⅛	48⅛	54⅛	66⅛	78⅛	48⅛	55⅞	67⅞	79⅞
B	2¼	2¼	2⅝	2⅝	2⅝	2⅝	3⅛	3⅛	3⅛	3⅛	3⅛	3⅛	3⅛	3½	3½	3½
C	30⅜	33⅜	41¼	47¼	32¼	35¼	42⅞	48⅞	33⅞	36⅞	42⅞	48⅞	33⅞	39⅛	45⅛	51⅛
D	29⅜	32⅜	41	47	32	35	42⅞	48⅞	33⅞	36⅞	42⅞	48⅞	33⅞	38¾	44¾	50¾
E	21	21	21	21	21	21	21	21	21	21	21	21	21	21	21	21
F	30	30	30	30	30	30	30	30	30	33	33	33	33	33	35	35
G	8⅛	8⅛	8⅛	8⅛	8⅛	8⅛	8⅛	8⅛	8⅛	8⅛	8⅛	8⅛	8⅛	8⅛	8⅛	8⅛
H	15¼	15¼	15¼	15¼	15¼	15¼	15¼	15¼	15¼	15¼	15¼	15¼	15¼	15¼	15¼	15¼
J	44⅛	50⅛	62⅝	74⅝	44⅝	50⅝	62⅝	74⅝	44⅝	50⅝	64⅝	76⅝	46⅝	52⅝	64⅝	76⅝
K	2 9/32	2 9/16	2 25/32	2 25/32	2 25/32	2 25/32	2 25/32	2 25/32	2 25/32	2 25/32	3⅛	3⅛	3⅛	3⅛	3⅛	3⅛
L	33⅛	39⅛	51⅛	63⅛	33⅛	39⅛	51⅛	63⅛	33⅛	39⅛	51⅛	63⅛	33⅛	39⅛	51⅛	63⅛
M	11¾	11¾	12	12	12	12	12	12	12	12	15	15	15	15	15	15
N	23	23	25	25	25	25	25	25	25	25	27	27	27	27	27	27
O	33	33	33	33	33	33	33	33	33	33	36	36	36	36	36	36
P	41⅛	47⅛	59⅛	71⅛	41⅛	47⅛	59⅛	71⅛	41⅛	47⅛	59⅛	71⅛	41⅛	47⅛	59⅛	71⅛
R	31⅝	31⅝	31⅝	31⅝	31⅝	31⅝	31⅝	31⅝	31⅝	31⅝	31⅝	31⅝	31⅝	31⅝	31⅝	31⅝
S	4	4	6	6	6	6	6	6	6	6	6	6	6	8	8	8
T	4	4	4	4	4	4	4	4	4	4	6	6	6	6	6	6
X	6	6	6	6	6	6	6	6	6	6	6	6	6	7	7	7
Z	6	6	6	6	6	6	6	6	6	6	6	6	6	6	6	6

For Material Weighing 100 Pounds per Cubic Foot. Dimensions in Inches

Length of Conveyor	0 to 25 ft. Centers				26 to 50 ft. Centers				51 to 75 ft. Centers				76 to 100 ft. Centers			
No. of Conveyor	2571	2591	2611	2631	2651	2671	2691	2711	2731	2751	2771	2791	2811	2831	2851	2871
A	46⅜	52⅜	64⅜	78⅛	46⅜	54⅛	66⅛	78⅛	48⅛	54⅛	67⅞	79⅞	49⅞	55⅞	67⅞	82⅛
B	2⅝	2⅝	2⅝	3⅛	2⅝	3⅛	3⅛	3⅛	3⅛	3⅛	3½	3½	3½	3½	3½	4⅛
C	32¼	35¼	41¼	48⅞	32¼	36⅞	42⅞	48⅞	33⅞	36⅞	45⅛	51⅛	36⅛	39⅛	45⅛	53⅝
D	32	35	41	48⅞	32	36⅞	42⅞	48⅞	33⅞	36⅞	44¾	50¾	35¾	38¾	44¾	53⅜
E	21	21	21	21	21	21	21	21	21	21	21	21	21	21	21	21
F	30	30	30	30	30	36	30	36	33	33	33	35	33	33	35	35
G	8⅛	8⅛	8⅛	8⅛	8⅛	8⅛	8⅛	8⅛	8⅛	8⅛	8⅛	8⅛	8⅛	8⅛	8⅛	8⅛
H	15¼	15¼	15¼	15¼	15¼	15¼	15¼	15¼	15¼	15¼	15¼	15¼	15¼	15¼	15¼	15¼
J	44⅝	50⅝	62⅝	76⅝	44⅝	52⅝	64⅝	76⅝	46⅝	52⅝	64⅝	76⅝	46⅝	52⅝	64⅝	77⅛
K	2 25/32	2 25/32	2 25/32	3⅛	2 25/32	3⅛	3⅛	3⅛	3⅛	3⅛	3⅛	3⅛	3⅛	3⅛	3⅛	4
L	33⅛	39⅛	51⅛	63⅛	33⅛	39⅛	51⅛	63⅛	33⅛	39⅛	51⅛	63⅛	33⅛	39⅛	51⅛	63⅝
M	12	12	12	15	12	15	15	15	15	15	15	15	15	15	15	22¼
N	25	25	25	27	25	27	27	27	27	27	27	27	27	27	27	39
O	33	33	33	36	33	36	36	36	36	36	36	36	36	36	36	43
P	41⅛	47⅛	59⅛	71⅛	41⅛	47⅛	59⅛	71⅛	41⅛	47⅛	59⅛	71⅛	41⅛	47⅛	59⅛	71⅛
R	31⅝	31⅝	31⅝	31⅝	31⅝	31⅝	31⅝	31⅝	31⅝	31⅝	31⅝	31⅝	31⅝	31⅝	31⅝	31⅝
S	6	6	6	6	6	6	6	6	6	6	8	8	8	8	8	8
T	4	4	4	6	4	6	6	6	6	6	6	6	6	6	6	6
X	6	6	6	6	6	6	6	6	6	6	7	7	7	7	7	8
Z	6	6	6	6	6	6	6	6	6	6	6	6	6	6	6	6

Pan Conveyors
and
Spiral Conveyors

Section

9

Jeffrey Cast Iron Pan Conveyor handling ashes in basement of boiler house. Ashes are drawn through Pit Doors into the Conveyor, which is built for hard, continuous service.

THE Cast Iron Pan is a modified form of the overlapping steel pan or steel apron in which the design is such as to withstand shrinkage strains due to the handling of hot materials. It has but one discharge point.

The Round Bottom Steel Pan is capable of carrying a much greater capacity for a given width and is especially adapted to the handling upon very steep inclines of materials which have a tendency to readily flow.

The Steel Overlapping Drop Pan Conveyor as shown below can be installed to deliver at various fixed points into Bins or Chutes.

Round Bottom Steel Pan Conveyor installed for carrying coal from railroad to storage bin over ovens in a large Coke Plant.

Overlapping Pan Conveyor handling coal from beneath a storage bin.

Overlapping Drop Pan Conveyor delivering coal to various fixed points.

Cast Iron Overlapping Type

THE Cast Iron Overlapping Pans are well adapted to the handling of Ashes and other similar abrasive or semi-abrasive material, as none of the material comes in contact with the moving parts.

Cast Iron Overlapping Pans.

Mounted between two strands of Steel Thimble Roller Chain, these pans form an endless moving trough.

Skirt Plates mounted as shown in the illustration above protect the chain from any spill in loading the Conveyor.

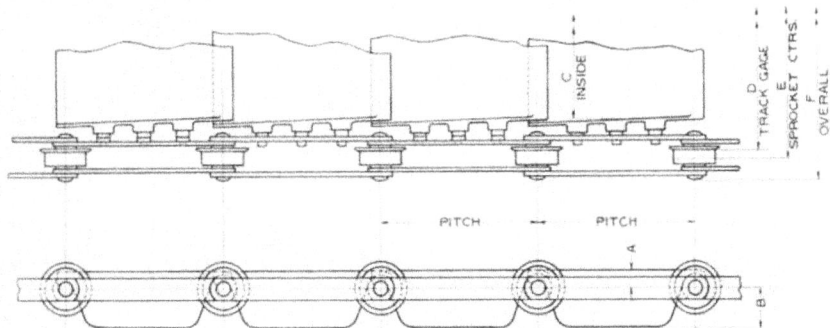

*Capacity in Cubic Feet per hour	*Capacity in Tons per hour	Chain		Weight per running ft. of Conveyor Lbs.	Dimensions—Inches					
		No.	Pitch		A	B	C	D	E	F
809	40	276	12	45	¾	2⁵⁄₁₆	12	16⁵⁄₁₆	17⁵⁄₁₆	20¾
1240	62	276	12	55	¾	2⁵⁄₁₆	18	22⁵⁄₁₆	23⁵⁄₁₆	26¾
1766	88	182½	18	77	2	2	17	22½	24¼	28¹³⁄₁₆
4000	200	182½	18	95	2	4⅜	23¼	30	31¾	36⁵⁄₁₆

*Capacities given for Conveyor Speed at 100 ft. per minute handling Material weighing 100 lbs. per cu. ft.

The above view shows a Jeffrey Spiral Conveyor
distributing coal over bunkers in a modern power
plant. This is ordinarily the first step in economy
beyond simply an elevator with gravity spouts.

At the left is shown the Boiler Room beneath the
bunkers illustrated above, with a Jeffrey Weigh
Larry delivering coal to stokers.

THE Spiral Conveyor is for those inaccessible
places which will not permit of a return
strand of conveyor such as immediately under
floors, directly under roofs or through shallow
roof trusses. The Spiral Conveyor has no return
of any kind and thus meets the requirements of
many power plants of such local conditions.

At the right the receiving end of the Spiral
Conveyor, pictured at top of page, is being
fed by a Jeffrey Standard Bucket Elevator.

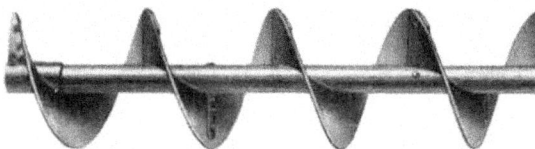

Sectional Flight Spiral Conveyor

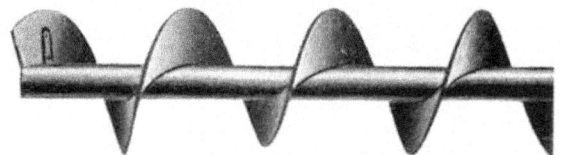

Steel Helicoid Continuous Flight Conveyor

SPIRAL Conveyors are made to carry loose bulk material which is not of a very gritty or sticky nature and of a maximum size not greater than one fourth the diameter of the spiral. The best service is rendered by a spiral conveyor handling lumpy, unsized or heavy abrasive materials such as Sand, Ashes, etc., when the depth of material does not exceed one-third the diameter of conveyor.

Ordinarily when handling non abrasive material of a size not greater than one-sixteenth the diameter of the conveyor, material should be fed uniformly so it will have a depth in the trough not greater than one-half the diameter of the Conveyor.

Grains and very light materials are often carried to a depth equal to the diameter of the spiral.

The Spiral Conveyor recommends itself for the handling of very fine and dusty materials as the trough can be fully enclosed. Ordinarily the Spiral Conveyor operates in a wood trough having a curved steel lining in the bottom, although a steel trough is often used.

The Spiral Conveyor can be fed at any point along its length and discharged at many places thru valves in the trough.

Spiral Conveyors should never be used to handle material likely to contain foreign substances such as scrap iron.

In selecting the size of a conveyor from the table below, it is always good practice and economy to use the next larger conveyor rather than to exceed "Sized Material" "Maximum Capacities" listed. In all cases the size of Conveyor should be governed by the maximum size piece rather than capacity. If then the capacity is greater than desired reduce the speed until the required capacity is reached. Do not run conveyors faster than necessary to obtain the capacity desired.

Dia. Conveyor Inches	Dia. Coupling Shaft Inches	Size of Pipe		Size Material		Light Non-Abrasive Material as Grain§			Heavy Non-Abrasive Material as Coal§			Heavy Abrasive Material as Sand and Ashes‡		
		Sectional Conveyor	Helicoid Conveyor	Max. Uniform Size **	Max. Un-sized †	Gauge Flights	Max Speed R.P.M.	Max. Cap'c'y Cu. Ft. Per Hr.	Gauge Flights	Max. Speed R.P.M.	Max. Cap'c'y Cu. Ft. Per Hr.	Gauge Flights	Max. Speed R.P.M.	Max. Cap'c'y Cu. Ft. Per Hr.
4	1	1	1¼	⅜	1	18	220	171	10	110	86	3/16	90	46
6	1½	1½	1¾	⅝	1½	16	200	528	10	100	264	3/16	80	138
9	1½	1½	2	¾	2¼	14	175	1659	10	85	806	3/16	70	405
9	2	2	2½	¾	2¼	14	175	1619	10	85	786	¼	70	405
10	1½	1½	2	⅞	2½	12	160	2096	10	80	1048	3/16	65	517
12	2	2	2½	1	3	12	150	3390	3/16	75	1695	¼	60	822
12	2 7/16	2½	3	1	3	12	150	3330	3/16	75	1665	¼	60	822
12	3	3	3½	1	3	12	150	3240	3/16	75	1620	5/16	60	822
14	2 7/16	2½	3	1⅛	3½	10	140	5010	3/16	70	2457	⅜	55	1199
16	3	3	3½	1⅜	4	10	130	6916	¼	65	3458	⅜	50	1630
16	3	4	4	1⅜	4	10	130	6685	¼	65	3341	⅜	50	1630
18	3	3	......	1½	4½	10	120	9180	¼	60	4590	⅜	45	2083
18	3	4	......	1½	4½	10	120	8900	¼	60	4590	⅜	45	2083
20	3	3	......	1¾	5	3/16	115	12155	¼	55	5813	⅜	45	2862
20	3	4	......	1¾	5	3/16	115	12155	¼	55	5813	⅜	45	2862

**About 90% of material of "Maximum Uniform Size" listed.

†Not more than 10% of the material to be of the "Maximum Unsized" listed.

††Capacities given are at maximum R. P. M. uniform and continuous flow of material for one hour. Other capacities directly proportional to speed. To maintain the listed capacities care must be taken that the quantities required can be fed to the conveyor under the operating conditions.

§Capacity figured with the depth of material equal to one-half diameter of conveyor.

‡Capacity figured with the depth of material equal to one-third diameter of conveyor.

When one conveyor discharges into another, the receiving conveyor, unless of larger diameter, should run 5 R. P. M. faster than the delivering conveyor and may exceed the maximum allowable speed by this amount.

The values given above are not given as specific rules but as guides in good general practice wherein there are acceptable variations depending upon the nature of the material handled, nature of the service, power consumption and the life of the conveyor.

For sticky materials consider use of Ribbon Conveyor. Information furnished upon request.

For wet gritty materials such as Ashes consider the use of Cast Iron Spiral Conveyor. Information furnished upon request.

Turning Spiral Conveyor end for end does not change it from one hand to the other but it does change the side of the flights working against the material.

Reversing the direction of rotation of a conveyor changes the direction in which the material travels.

Conveyors should operate with lugs on side opposite to the one in contact with the material.

Maximum angle of inclination with standard pitch 30 degrees.

Horse-power required for Spiral Conveyors.

$$H. P. = \frac{F C L W}{33000}$$

C = Capacity of Conveyor in cu. ft. per minute.

L = Length of Conveyor in feet.

W = Weight of material in pounds per cu. ft.

F = 1.3 for light non-abrasive materials such as grain.

    2.5 for heavy non-abrasive materials such as coal, cement, etc.

    4.0 for heavy abrasive materials such as Sand, Ashes, etc.

The power required to drive a Spiral Conveyor depends entirely upon the nature of material handled. Therefore the above formula can be only approximately correct.

Skip Hoists

Section
10

THE design of the Jeffrey Skip Hoist especially adapts it to the handling of ashes and similar abrasive materials, as any material which may be handled does not come in contact with the operating mechanism.

The Skip Hoist is unquestionably the most economical means of handling ashes in either small or large boiler houses. It is simple in construction; of low initial cost; and inexpensive to operate as it is in service only when actually carrying a load.

The wide-open-mouth of the Skip permits the carrying of large clinkers without previous crushing.

The Jeffrey Skip Hoist is made in two Standard sizes using a 27 cubic foot and 40 cubic foot bucket, see pages 182 and 183.

Where conditions will not permit of a vertical skip hoist the runways may be inclined as in the accompanying views. Guard rails, as shown in upper illustration on opposite page, are omitted when the angle of runway exceeds 10 degrees from vertical.

Head Frame

THE Head Frame of the Jeffrey Standard Skip Hoist is designed to serve any type of Storage Bin, and can be joined to either a vertical or inclined runway.

The operation of the Skip Hoist is automatic in every respect and therefore requires little attention. When the Skip Car reaches the dumping point it automatically opens the door of the storage bin and upon discharging its load, starts back, closing the door simultaneously, thus eliminating any spill and scattering of dust, or water entering the bin.

While it is not necessary to house in the head frame, it is so recommended, as it affords protection from the weather, and thereby increases its life.

The Foot End

The Skip car comes to rest in its lower position slightly below the floor or ground level and material can be discharged into it either by wheelbarrow or ash dump car. Where large capacities are to be handled, requiring almost continuous service of the Skip Hoist, an automatic feeder in connection with hopper can be installed to load car, which in turn will operate the Hoist.

The Ash Dump Car

The Ash Dump Car consists of a steel hopper, pivoted on a short wheel base truck which will operate upon not less than a 15 foot radius curve. The wheels are chilled iron and provided with roller bearings which insure a long life and easy operation. On account of the absence of doors or locks, this car can be dumped and returned to the loading point very quickly.

Ash Dump Car.

Skip Car

Skip Cars

Both sizes of Skip Cars are similarly designed, the sides of which are constructed with heavy steel plate, with bottom of greater thickness, all securely held with closely spaced hot pressed rivets. Cars are mounted on 8-inch diameter cast iron wheels and provided with bale of ample strength to withstand severe service. The bale has double channel cross beams at the top with connections for the operating steel rope.

Counterweight

AS in the case of the cars, the counterweights are alike, except in size so as to equal the weight of the car and half the full loading, thereby requiring a hoisting engine of minimum size by having to raise but half of the load in the car when going up, and lifting an equal amount of the counterweight load in coming down empty. These counterweights consist of steel forms which are filled with concrete by the purchaser. They are mounted upon cast iron carrying rollers which operate on the back of the runway channels.

Control

The Hoisting Engine is of the worm and gear type and is electrically driven. The motor used being especially built to withstand the hard usage of elevator service, with the starting torque greatly exceeding the running torque. This allows the motor to start at full load on almost the same current that is required to run, thereby making it very economical in operation.

Skip Car Counterweight

The worm and worm shaft are forged in one piece, accurately machined and equipped with thrust bearings. The worm wheel is composed of a cast iron center with a renewable bronze rim, the teeth being machine cut. They are inclosed within a cast iron dust tight casing and run in oil, insuring lubrication of the gears and thrust bearings. The worm wheel and drum are cast together on the drum shaft, thus doing away with all keys and set screws. The grooves in the drum are turned to properly fit the cable, insuring long wear.

A self-adjusting brake is provided which applies pressure to the brake wheel at all points around its circumference, thus eliminating the jerking, jarring and chattering quite common to this service. The brake is held in the released position electrically while the engine is in operation and is applied automatically upon the breaking of the circuit. In this way there is positive assurance that the car cannot run back in case of accident to the power lines.

The Skip Hoist is automatic in operation—by pressing a button located in a convenient place at the bottom, the up-direction switches are closed electrically, starting the car which runs to the top. Passing over a hatch switch the car is brought to a stop in a dumping position. A time relay holds it in this position for a short interval of time to allow for the complete discharge of the load. The down-going switches are then automatically closed, starting the car down. When the car reaches the lower position, the counterweight passes onto a shoe, breaking the current by means of a hatch switch which brings the car to rest in the loading position.

Electric Traction Hoist

By this controlling device, much time is saved as the car is back in the loading position by the time the operator has returned with another load of material.

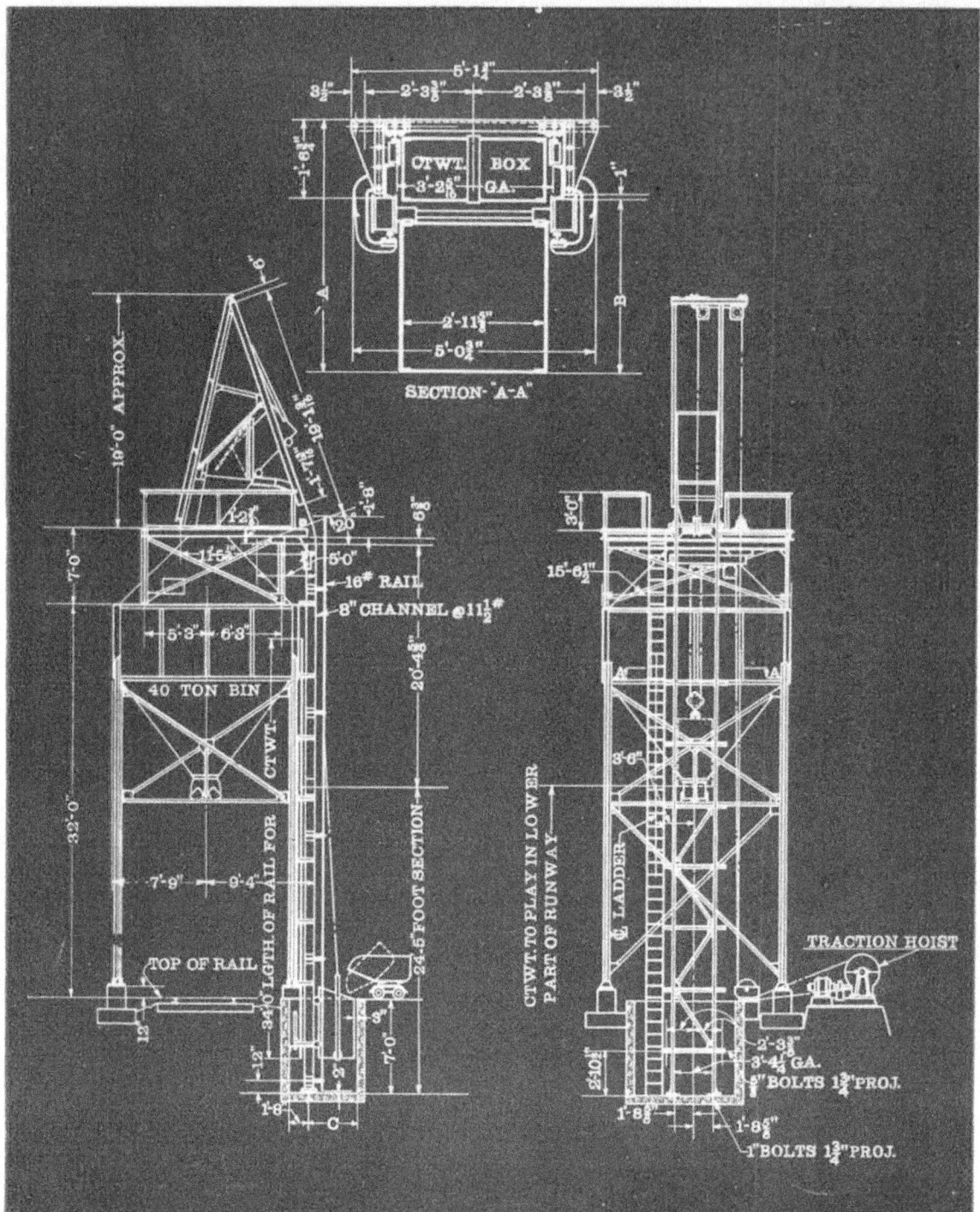

General Layout of Skip Hoist with 40 Ton Ash Bin

Capacity of Bucket	Electric Hoist			Approx. Speed in Feet per Min.	Dimensions		
	Number	Weight	Unbalanced Load		A	B	C
27 Cubic Feet	4	2000	1200	75	4'-8½"	3'-1¾"	3'-8¾"
40 Cubic Feet	5	3500	2400	75	5'-1⅛"	3'-6⅜"	4'-1⅜"

⅝" Crucible Steel Hoisting Rope, used on both sizes.

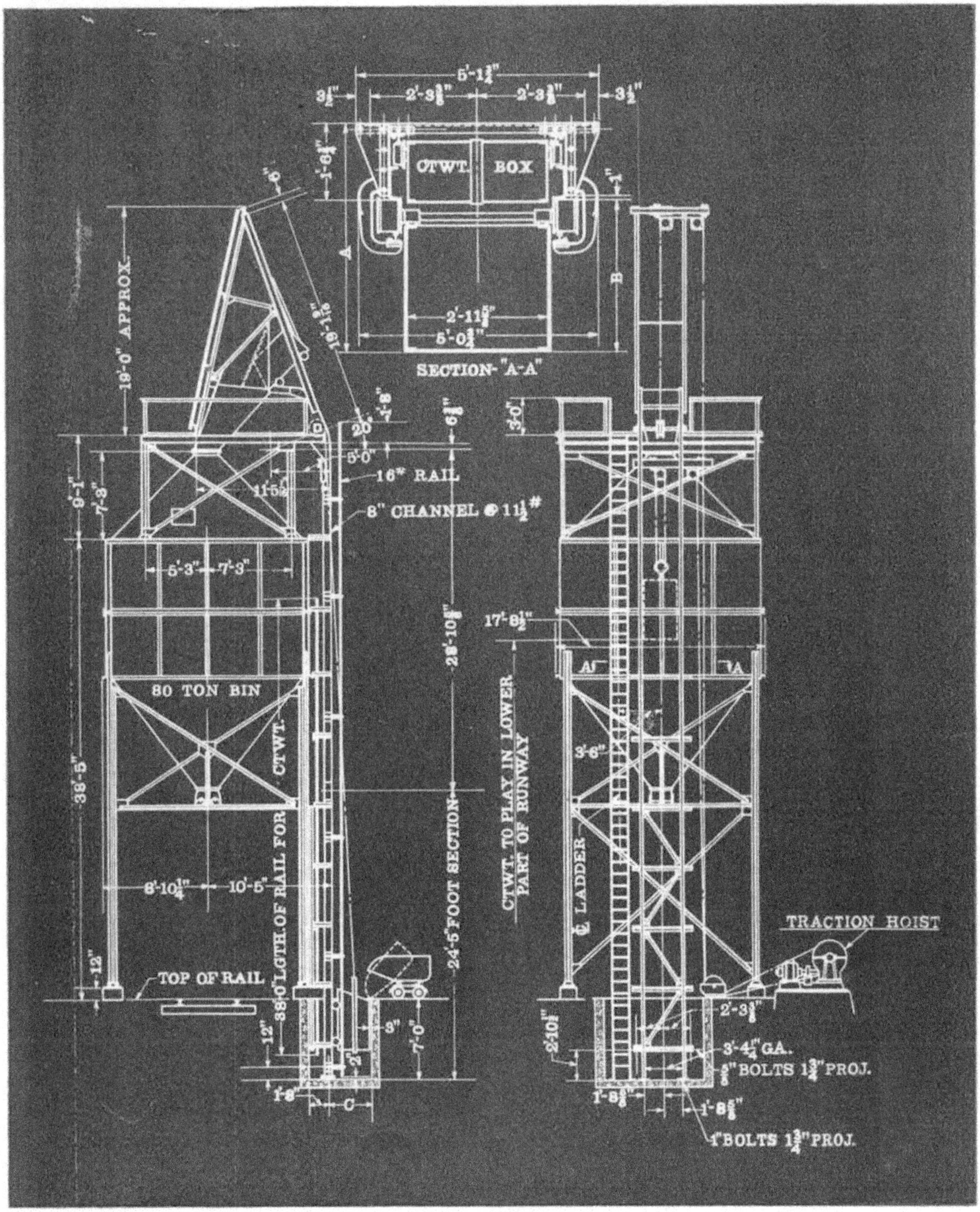
CTWT. BOX
SECTION-"A-A"
19'-0" APPROX.
80 TON BIN
16# RAIL
8" CHANNEL @ 11½#
TOP OF RAIL
38'-0" LGTH. OF RAIL FOR CTWT.
24'-5" FOOT SECTION
28'-10⅝"
CTWT. TO PLAY IN LOWER PART OF RUNWAY
℄ LADDER
TRACTION HOIST
3'-4¼" GA.
⅝" BOLTS 1¾" PROJ.
1" BOLTS 1¾" PROJ.

General Layout of Skip Hoist with 80 Ton Ash Bin

Capacity of Bucket	Electric Hoist			Approx. Speed in Feet per Min.	Dimensions		
	Number	Weight	Unbalanced Load		A	B	C
27 Cubic Feet	4	2000	1200	75	4'-8½"	3'-1¾"	3'-8¾"
40 Cubic Feet	5	3500	2400	75	5'-1⅛"	3'-6⅜"	4'-1⅜"

⅝" Crucible Steel Hoisting Rope, used on both sizes.

Delivering Ashes to Railroad Car from Power House of Industrial Plant.

Rear View Showing Loaded Car Ascending to Top of Storage Bin.

FOR the sake of convenience and economy, it is often advisable to have Ash Handling Equipment entirely independent from that of the Coal Handling. When delivering coal to bunkers at times it becomes necessary to remove the ashes, which would cause inconvenience if the same equipment were used for both.

Where the usage is intermittent, such as the handling of ashes from boiler house to storage bin, the Skip Hoist is the most satisfactory equipment, regardless of whether it be a small power house or a large one. In the majority of cases an ash dump car, illustrated on page 180, is used to carry the ashes to Skip car.

Crushers

Section
11

**A Crusher for
Any Size Lump**

18″ x 18″ Crusher
8″ Maximum Lump
Small Capacities

24″ x 24″ Crusher
Average Conditions
14″ Maximum Lump—50 Tons per Hour

30″ x 30″—36″ x 36″
36″ x 54″ Crushers
Big Lumps
Large Capacities

Reduces Lump Coal to Any Desired Size in a Single Operation

Patented

THE Jeffrey Single Roll Crusher is built in five sizes, as given above. The 18″ x 18″ which takes a maximum lump of 8″ is used for small capacities. The 24″ x 24″ Crusher meets average requirements such as Boiler Plants, Gas Producer Plants and Pulverized Coal Plants. It takes a maximum lump of 14″ and has a capacity of 50 tons per hour.

The larger Single Roll Crushers, 30″ x 30″ 36″ x 36″ and 36″ x 54″ take run-of-mine coal and have large capacities. These machines are generally for service at the mine in reducing the entire output to a size suitable for use with mechanical stokers either in boiler house or locomotive service.

**Cross-section view of Jeffrey Single Roll Crusher
illustrating how it reduces lump coal in a single operation**

Such materials as Salt, Bone and Alum are also being successfully reduced by Jeffrey Single Roll Crushers.

Here is shown a uniform product of 1¼″ and under for automatic stokers, and 6″ and under for hand firing, both obtainable with the Jeffrey Single Roll Crusher.

1¼″ **and under for Automatic Stokers**

6″ **and under for Hand Firing**

Typical Installations of Jeffrey Single Roll Crushers

Crusher in operation in Power House of a plant manufacturing sewing machines. Note uniformity and fineness of crushed coal in this Jeffrey Steel Apron Conveyor.

This Crusher installed in a Refining Company is driven by a steam engine. An Apron Conveyor operating from under a track hopper feeds the crusher, which in turn discharges into a Pivoted Bucket Conveyor.

Crusher installed in a large Central Station. When reduced, coal passes thru a Spiral Conveyor to Bucket Elevator which discharges the Coal into the bunkers.

An installation in the Tipple Building of a coal mine for the production of fine coal.

A Crusher installed for grab bucket feed in the plant of a Street Railway Company.

Jeffrey Single Roll Crusher meets the requirements of the Railroad Coaling Station.

Construction and Operation

Patented

Front view of Jeffrey Single Roll Crusher, showing Cut Tooth Steel Gears and Safety Guard.

3491-C

THE construction of the Jeffrey Single Roll Crusher is very rigid, and it will stand the most severe service. Our design may be called almost brutal, since the care the crushers receive and the use to which they are put calls for more brute strength and endurance than for any over-refinement of parts. And yet, with much care, these machines have been well proportioned.

The design is extremely simple, consisting of a heavy cast iron frame in which are mounted a crushing roll and a breaker plate. The breaker plate is hinged at its upper end and is held in position by a pair of adjustable tension rods at the lower edge by means of which the clear opening between the breaker plate shoe and the surface of the roll can be varied to give any product required.

A clamping effect is produced by the proper adjustment of the cross-rod bolts between the side frames, whereby sufficient friction is brought upon the hinged breaker plate to eliminate chattering and to assist the safety device.

The frame is of the box type section, very stiff and rigid. All joints are machined, as also are the teeth of the heavy steel gears. All parts are made to jig so that repairs can readily be furnished. The Main or Roll Bearings are equipped with readily renewable bronze liners of special composition, while the countershaft bearings are supplied with renewable die cast babbitted bushings. Lubrication is obtained through large compression grease cups. Because of the speed of operation, this crusher is especially adapted for electric motor drive; a belt from the motor pulley to the band wheel on the crusher being usually all that is required. When space is very limited, the pulleys and belt are replaced by a pair of gears, having the same safety device.

Roll Shaft and Countershaft of large diameter—Roll Shaft Bearings have renewable bronze bushings. Countershaft bearings have renewable babbitt bushings—Grease cups provide lubrication.

Toothed Segments

A close-up view of the Crushing Roll, showing the Long Feeder Teeth which crush large lumps and the Short Crushing Teeth which give Uniform Product.

NARROW slots in the shoe of the breaker plate enable the long teeth to pass without dragging oversize pieces with them. These teeth not only act as feeders but they positively grip the large pieces and break them up in sizes which readily and unhesitatingly enter down deeper into the maw of the machine.

By making the smaller teeth on the segments of the peculiar shape shown, the proper reduction is made with a minimum amount of slack. The toothed segments are usually made in our special hard manganoid metal. The long teeth are made of hardened, drop forged steel and inserted into the body of the segment.

Breaker Plate with Renewable Shoe

Toothed segments are bolted to convex surface of the drum so as to completely cover it. This forms a very durable and satisfactory crushing surface. The frame and hopper are so arranged that by removing the light steel guard plate and hand hole cover plates access may be had to the bolts and the segments removed and replaced by new ones without disturbing either the roll or the hopper. This is very convenient when crusher is installed in connection with a large hopper or complicated chute.

Accessibility of Segment Bolts.

Safety Device

AS a protection against such foreign material as shown here, of which a surprisingly large percentage comes in a car of coal, the Driving Pulley of the Single Roll Crusher is not keyed to the shaft, but is mounted on a separate hub which it drives through a set of wood pins inserted in holes in the pulley arms. When any undue strain comes on the machine from any cause, these wood pins shear off, and the crushing roll stops, while the pulley keeps on revolving. After the cause of the trouble is removed, new wood pins again put the machine in operative condition.

A pair of heavy springs are placed on the tension rods. These springs do not move under ordinary working conditions, but when an undue pressure comes on the breaker plate, they act as a cushion, giving way slightly, taking up the inertia of the parts and allowing time for the pins to shear without breaking more important elements in the machine.

Portable Crusher for Grab Bucket Feed

Portable Single Roll Crusher with Large Hopper for Grab Bucket Service. Made to travel as loader over conveyor along side of railroad cars, or directly over storage bins.

Portable Single Roll Crusher with motor and gear cover removed to show the rigid and compact yet very accessible construction of this highly efficient crushing unit.

Dimensions and Approximate Weights of Portable Crushers complete, except Motor.

Size Crusher	Weight Lbs.	Dimensions				
		A	B	C	D	E
18″ x 18″	6000	5′- 0″	6′- 0″	7′- 0″	6′-4″	8′- 1″
24″ x 24″	10500	5′- 2″	6′- 2″	7′- 2″	7′-0″	9′- 6″
30″ x 30″	15500	5′- 9″	6′- 9″	7′- 9″	6′-8″	10′-11″

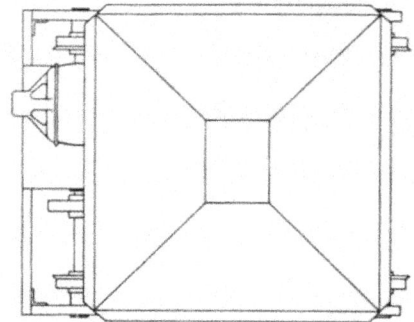

General dimensions of the Standard Single Roll Crusher assembled as a portable machine. Change in dimensions of hopper made to suit requirements.

Line drawings of other portable sizes furnished on request.

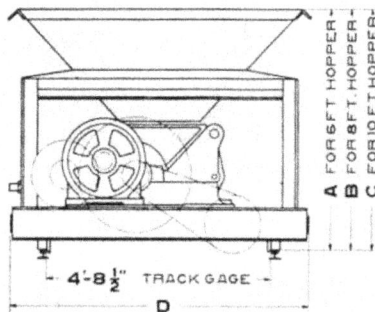

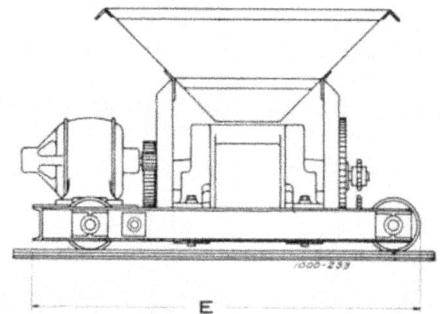

TABLE I—FOR HARD BITUMINOUS COAL

TONS PER HOUR	1"	1¼"	1½"	2"	3"	4"	5"	6"
10								
15	18" x 18" Crusher							
20			8" Cubes Maximum Feed					
25				For Larger Lumps				
30					Use Larger Crusher			
40	24" x 24"							
50	Crusher							
60			14" Cubes					
75	30" x 30"			Maximum				
100	Crusher							
125			20" Cubes Maximum					
150	36" x 36"							
175	Crusher							
200								
250	36" x 54"							
300	Crusher							
350								
400								
450	For							
500	These Capacities							
550	and Products							
600	Use More							
650	Than One							
700	Machine							

Capacities of Single Roll Crushers

Using HARD Bituminous Coal

such as Indiana Block, West Virginia Splint, Illinois, Iowa, Colorado, Wyoming, Penn Freeport, Kittanning and Cannell.

For Methods of using Tables see Example of SOFT COAL on the following page.

NOTE—With Table I use Horsepower Table IV page 193.

Using MEDIUM HARD Bituminous Coal

such as W. Virginia Thacker, Panther, Banner, Pittsburgh No. 8, Coalburg, Kentucky, Harlan, Hazard No. 4, No. 7 Block and Ohio Hocking.

*By "Size of Product" is meant average results, 80 to 90 per cent. pass screen indicated. To increase capacities of Single Roll Crusher see Page 193.

NOTE—With Table II use Horsepower Table V on page 193.

TABLE II—FOR MEDIUM HARD BITUMINOUS COAL

TONS PER HOUR	1"	1¼"	1½"	2"	3"	4"	5"	6"
10								
15			18" x 18"					
20			Crusher					
25					10" Maximum Feed			
30					For Larger Lumps			
40					Use Larger Crusher			
50								
60	24" x 24"							
75		Crusher						
100			16" Cubes					
125	30" x 30"			Maximum				
150	Crusher							
175			22" Cubes					
200	36" x 36"		Maximum					
250								
300			Crusher					
350								
400	36" x 54"							
450								
500			Crusher					
550								
600	For these							
700	Capacities							
800	and Products Use							
900	More Than One Machine							

TABLE III—FOR SOFT BITUMINOUS COAL

TONS PER HOUR	Size of Product*							
	1"	1¼"	1½"	2"	3"	4"	5"	6"
10								
15								
20								
25		18" x 18" Crusher						
30								
40			12" Cubes Maximum Feed					
50			For Larger Lumps					
60			Use Larger Crusher					
75		24"						
100		x						
125			24"					
150		30"		Crusher				
175		x			18" Cubes			
200			30"		Maximum			
250				Crusher				
300		36"			24" Cubes			
350		x			Maximum			
400			36"					
450								
500		36"						
550		x		Crusher				
600								
700	For These		54"					
800	Capacities							
900	and Products		Crusher					
1000	Use More Than							
1100	One Machine							

Capacities of Single Roll Crushers

Using SOFT Bituminous Coal

such as Pocahontas, Connellsville, New River, Pittsburgh Coking, and Youghiogheny.

*By "Size of Product" is meant average results, 80 to 90 per cent. pass screen indicated. To increase capacities of Single Roll Crusher, see page 193.

NOTE—With Table III use Horsepower Table VI on page 193.

Examples using the above Table for Soft Bituminous Coal:—

40 Tons per hour reduced to 1½" size of product and 12" maximum cubes feed calls for 18" x 18" Crusher.

250 Tons per hour reduced to 4" size of product and 24" maximum cubes feed calls for 30" x 30" Crusher.

300 Tons per hour reduced to 1¼" size of product and 24" maximum cubes feed calls for 36" x 36" Crusher.

Horse Power Required for Single Roll Crushers

TABLE IV

To be used with Table I, page 191

For Hard Bituminous Coal

Size of Product	1" 1¼"	1½" 2" 3"	4" 5" 6"	
Size of Crusher Selected from page 191	HORSE POWER For Each Ton of Coal Crushed Per Hour			Size of Motor to be Not less Than
18x18	½	⅓	⅙	7½ H. P.
24x24	½	⅓	⅙	10 H. P.
30x30	½	⅓	⅙	15 H. P.
36x36	½	⅓	⅙	20 H. P.
36x54	½	⅓	⅙	25 H. P.

TABLE V

To be used with Table II, page 191

For Medium Hard Bituminous Coal

Size of Product	1" 1¼"	1½" 2" 3"	4" 5" 6"	
Size of Crusher Selected from page 191	HORSE POWER For Each Ton of Coal Crushed Per Hour			Size of Motor to be Not less Than
18x18	⅓	²⁄₉	⅑	7½ H. P.
24x24	⅓	²⁄₉	⅑	10 H. P.
30x30	⅓	²⁄₉	⅑	15 H. P.
36x36	⅓	²⁄₉	⅑	20 H. P.
36x54	⅓	²⁄₉	⅑	25 H. P.

Example under Table IV:—Horsepower required to reduce 100 tons per hour to 1"-1¼" in a 30 x 30 Crusher is 100 times ½ (in Table) equals 50 H. P.

TABLE VI

To be used with Table III, page 192

For Soft Bituminous Coal

Size of Product	1" 1¼"	1½" 2" 3"	4" 5" 6"	
Size of Crusher Selected from page 192	HORSE POWER For Each Ton of Coal Crushed Per Hour			Size of Motor to be Not less Than
18x18	¼	⅙	¹⁄₁₂	7½ H. P.
24x24	¼	⅙	¹⁄₁₂	10 H. P.
30x30	¼	⅙	¹⁄₁₂	15 H. P.
36x36	¼	⅙	¹⁄₁₂	20 H. P.
36x54	¼	⅙	¹⁄₁₂	25 H. P.

To Increase Capacities

of the

Single Roll Crusher

The Capacity Tables, pages 191 and 192 are based on "STANDARD SPEEDS" of Rolls in Table below. These capacities however, may be increased or decreased 50% by a corresponding increase or decrease in the speed of crusher roll and also a corresponding increase or decrease in the Horse-Power of the motor required—with size of Motor in no case to be less than listed in Tables.

Speeds, Shipping Weights, Etc.

Size of Crusher	Standard Speed of Roll Rev. per Min.	Max. Speed of Roll Rev. per Min.	Approx., Shipping Weight Pounds	Floor Space See also Page 194	Size of Pulley—Inches	
					Diam.	Face
18x18	75	125	3200	5' 0" x 4'4"	34	6½
24x24	60	100	6500	6'11" x 5'4"	42	8½
30x30	50	75	10500	7'11" x 6'9"	48	10½
36x36	40	60	20000	8' 6" x 8'9"	60	15½
36x54	40	60	31000	9' 0" x 11'6"	66	19

General Dimensions of Crusher for Belt Drive

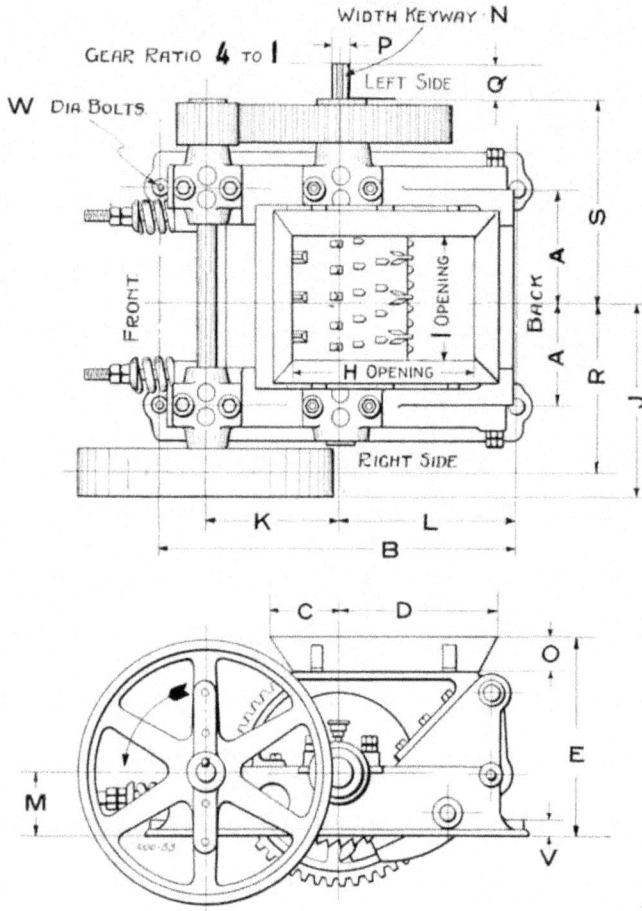

When ordering Crusher be sure to state on which side driving pulley is wanted. Crusher will be shipped assembled as shown, unless otherwise ordered.

If Crusher is to be Gear Driven give Motor Shaft Diameter, Keyseating, Shaft Extension and Speed

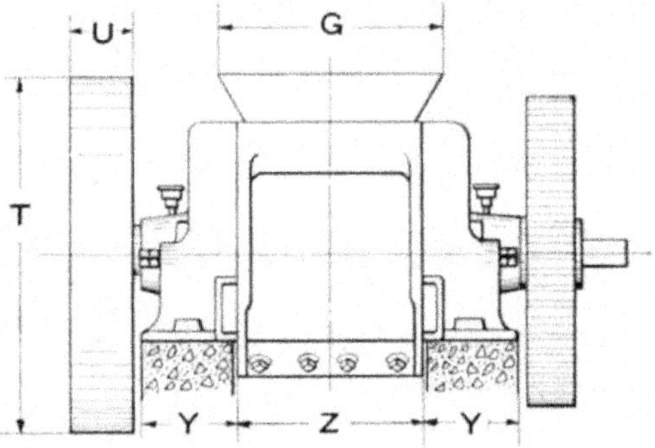

Table of Dimensions—Inches

Size Crusher	A	B	C	D	E	G	H	I	J	K	L	M
18" x 18"	14⅛	47	9¼	20¾	25	22	24	16	26½	17¾	23	8
24" x 24"	17⅞	55½	12½	23½	32½	30	27½	22	33	21	26	10½
30" x 30"	22⅛	65	14	27	37	36	32½	28	39¾	25½	30¾	12
36" x 36"	27¼	76	16¾	31¼	44¾	43½	40	35½	52¾	30½	37	15
36" x 54"	37½	84	15½	27½	47	59	38	51	68	34	38	17

Table of Dimensions—Continued

Size Crusher	N	O	P	Q	R	S	T	U	V	W	Y	Z
18" x 18"	9/16 x 9/32	4½	2 3/16	4½	23⅛	25½	34	6½	2	1	9½	19
24" x 24"	9/16 x 9/32	6	2 3/16	4¼	28¾	30¾	42	8½	2¼	1¼	11	25
30" x 30"	7/8 x 7/16	6	3 7/16	4¾	35⅜	36½	48	10½	3	1¼	13½	31
36" x 36"	1 x ½	6	3 15/16	5½	55	45	60	15½	3½	2	17½	37
36" x 54"	1 x ½	6	3 15/16	8	58½	58½	66	19	4	2	20	55

Tables give approximate dimensions only. Certified print for your installation furnished upon application.

Coal Storage
Equipment

Section
12

An Economical Method of Storing and Reclaiming
Large Quantities of Coal

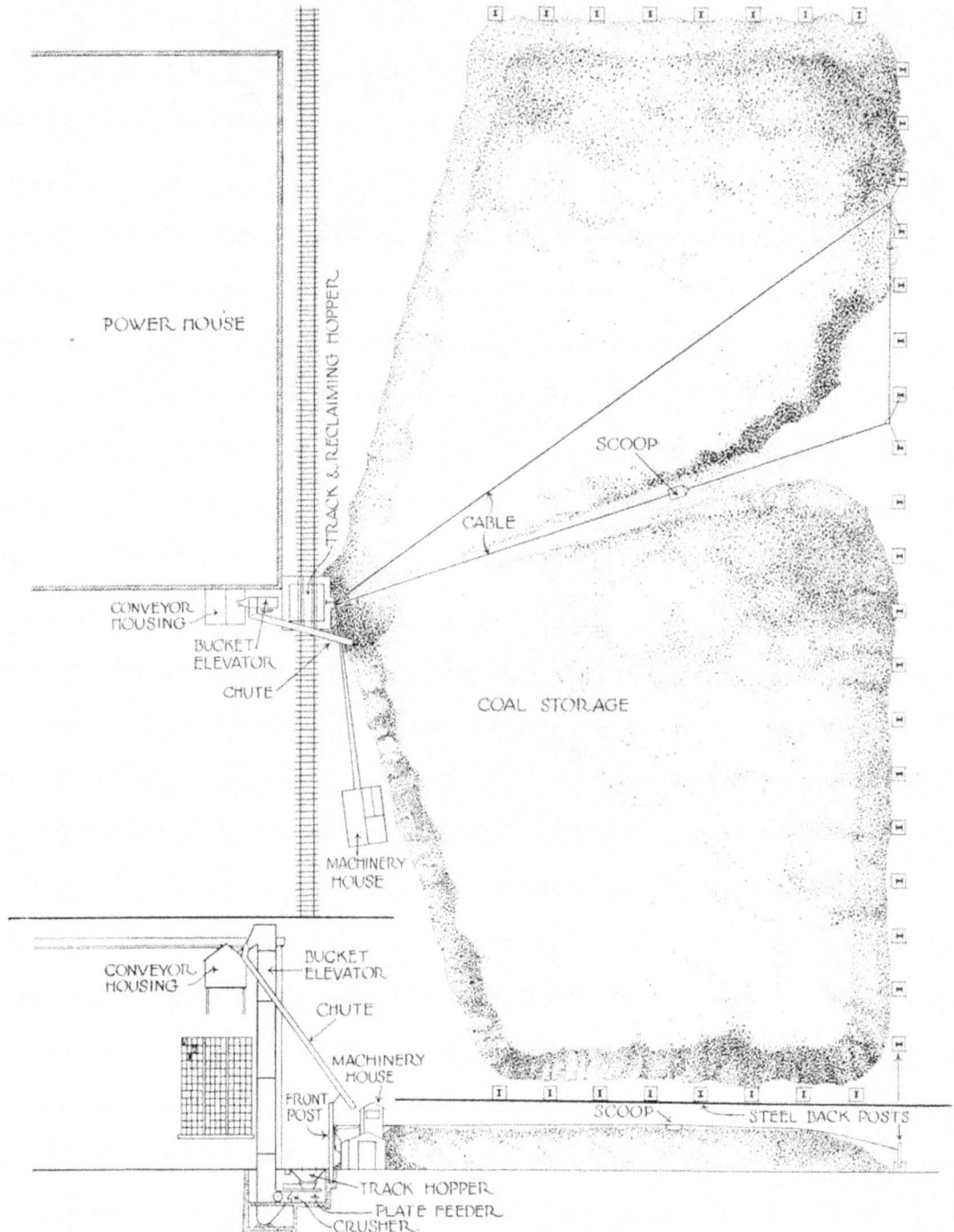

POWER HOUSE

TRACK & RECLAIMING HOPPER

SCOOP

CABLE

CONVEYOR HOUSING

BUCKET ELEVATOR

CHUTE

COAL STORAGE

MACHINERY HOUSE

CONVEYOR HOUSING

BUCKET ELEVATOR

CHUTE

MACHINERY HOUSE

FRONT POST

SCOOP

STEEL BACK POSTS

TRACK HOPPER

PLATE FEEDER

CRUSHER

THE Jeffrey Cable Drag Scraper affords the most economical and efficient means of storing and reclaiming large quantities of coal where an outside reserve storage must be maintained. This system can be arranged to suit any shape of storage area or capacity, as well as for separately storing several sizes or grades of coal.

The Ideal Arrangement

THE diagram on the opposite page shows what is considered an ideal arrangement for the handling and storing of large quantities of coal to insure the maximum of results. As shown, coal is received from railroad cars and discharged into track-hopper from where it is fed by plate feeder to Jeffrey Single Roll Crusher, discharging into a standard Bucket Elevator.

When storing, the coal is discharged by elevator into the path of scraper, and dragged out to the point desired. Since it is neither advisable nor practical to keep the reserve storage system active, the elevator is equipped with a two-way spout to allow coal to be delivered direct to a distributing conveyor operating over bunkers inside of power house. This eliminates the necessity of sending the coal through the storage system before it goes to the boiler house.

A Simple System Requiring But Little Machinery

In general, the Jeffrey Cable Drag Scraper System is remarkably simple in design and operation, consisting of a series of steel posts spaced at intervals to which the cables operating the Scraper are attached. Storing or reclaiming of the coal from any portion of the yard is accomplished by changing the Cable Tail Blocks from one post to another.

The Scraper, which consists of a steel bucket made in various sizes, fills itself and scrapes the coal to any desired point in the storage space. In reclaiming, the scraper is simply reversed on the cable and the coal scraped back to the reclaiming hopper, again passing through the crusher and elevator as explained above. Since the period for reclaiming is usually in the winter months, passing the coal through the crusher is advisable as it prevents frozen coal from causing damage to the elevator or distributing conveyor.

Drive and Control

The hoist is the double drum, direct geared cut toothed type, electrically operated. The drums are independently equipped with cone friction clutches. The driving gears are always in mesh, the driving drums operating in the same direction when engaged. All control levers are operated through quadrants and conveniently located in machinery house so one man can efficiently control the operation of the scraper and handle the entire storage system.

Detailed information on a Jeffrey Cable Drag Scraper System to suit your requirements will be furnished upon request.

Handling Coal from ground storage to motor truck with a Jeffrey Portable Belt Conveyor.

THE Jeffrey Portable Belt Conveyor, made in lengths of 18 feet to 60 feet and driven by either gasoline engine or electric motor, is designed to meet the needs of small Power Plants and Manufacturing Plants, in loading or storing material. It is a light and inexpensive machine, and sturdily built. Capacity varies from 20 to 50 tons per hour, depending upon the kind of material handled and the method used in loading the conveyor. It will handle all that two men can shovel onto it.

A close-up of the 60 ft. Jeffrey Portable Belt Conveyor shown in the lower illustration on opposite page.

The Jeffrey Portable Car Unloader is designed to unload Coal from hopper-bottom railroad cars directly into motor truck, or can be extended to storage pile. The Jeffrey Portable Car Unloader is built to fit in between rails and car hopper-door, or where a permanent installation is desirable, it can be placed beneath rails as shown in illustration.

Small Manufacturing Plants which receive large quantities of coal in hopper bottom railroad cars will find the Jeffrey Portable Car Unloader a great factor in reducing unloading costs and eliminating demurrage charges. Only one man is required to operate the Car Unloader, all operating levers being conveniently located on one side of the machine. Power is supplied by a 15 H. P., 4 cylinder gasoline engine, or an electric motor if desired.

THE Jeffrey Portable Scraper Conveyor is designed especially for loading domestic size bituminous coal into trucks and wagons. This machine is capable of handling one ton per minute. It is entirely self-contained and the discharge end is adjustable to take care of different heights of trucks. The discharge end may also be fitted with screen chute to separate the dust from the coal. Drive is furnished by an electric motor or gasoline engine.

Type "G" Radial Loader reclaiming coal from ground storage in the yards of a Retail Coal Company. The Jeffrey Radial Loader is adapted to the handling of Coal, Cinders, Sand, Crushed Stone and similar loose materials. Can also be furnished mounted on caterpillar tread. Capacity, 1 to 2 cubic yards per minute.

Large power plants using their ashes for filling purposes have found Jeffrey Storage Battery Locomotives, with a string of ash cars, the most economical and efficient means of conveying the ashes from the storage hoppers to the dump.

The battery capacity of the Jeffrey Storage Battery Locomotive is 63 A-10 Edison cells having a kilowatt hour capacity of 28 K. W. hours. It is capable of hauling 240 ton miles on level tracks on a single charge. Installation shows locomotive used by a large Public Service Corporation.

A JEFFREY Storage Battery Locomotive with a trip of six or eight cars in connection with an inclined trestle is used extensively for delivering ashes from ash pits directly into railroad cars where the daily volume of ashes is beyond the capacity of the ordinary ash bin.

Jeffrey 6-ton Electric Trolley Locomotive, with canopy, in quarry service

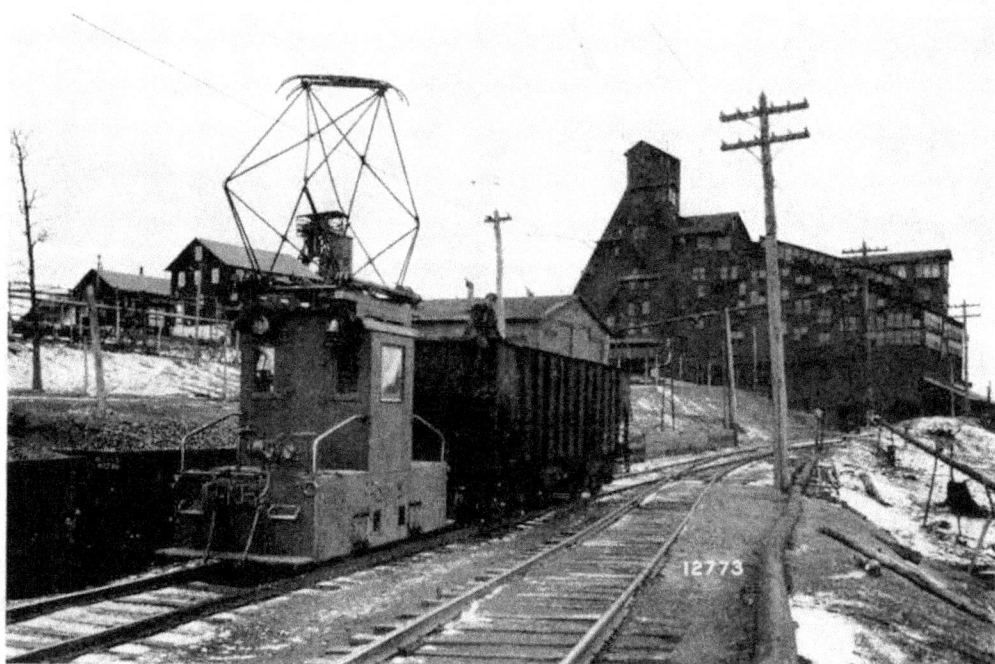

Jeffrey 20-ton Cab Locomotive used for handling railroad cars at a large anthracite coal mine

Mechanical Draft Fans

Section

13

Advantages of Mechanical Draft in Power House Service

POWER HOUSES are constantly looking for power and labor saving apparatus. They are daily striving to discover means by which operating expenses may be decreased and plant efficiency increased.

The economy of Mechanical Draft is becoming more and more apparent to the man who produces his own power.

Also there are innumerable cases throughout the industrial world where the Power House is desirous of increasing its Boiler Capacity and is unable to do so because of the limited Draft to be obtained from their present stack. Enlarging the stack is costly and difficult. The answer will be found in Mechanical Draft. Enlarged Boiler capacity, Draft Control, independence of weather conditions and the ability to burn cheaper fuel are made possible by the installation of Mechanical Draft.

It is highly important that the Combustion Engineer be very careful in specifying the volume of air and the pressure it is to work against when contemplating the use of Mechanical Draft.

The following pages contain enough data to enable him to be specific in making known to us his requirements.

Below is a Typical Example in Forced Draft to Find the Air Required

Given a battery of 1000 H. P. Boilers with ordinary grates running at rated capacity, steam pressure 125 lbs., temperature of feed water being 212 degrees F.

What is the volume required in cubic feet per minute when the outside temperature is 70 degrees F?

The total heat in steam at 125 lbs. pressure equals 1192.2 B.T.U., and total heat of the feed water at 212 degrees F. equals 180 B.T.U. and therefore, the total heat required to evaporate 1 lb. of water equals:

$$1192.2 \text{ B. T. U.}$$
$$-\ 180.0 \text{ B. T. U.}$$
$$\overline{\qquad\qquad}$$
$$1012.2 \text{ B. T. U.}$$

Assuming analysis of the coal consumed shows 12,000 B. T. U. per pound, and at a boiler efficiency of 70%, we find that: 12,000 x .70 ÷ 1012.2 = 8.3, or the pounds of water evaporated per pound of coal burned.

We know that one Boiler Horse-Power = 33,479 B. T. U., and so 33,479 ÷ 1012.2 = 33.1, or the pounds of water per boiler horse-power per hour.

If it requires 33.1 lbs. water per B. H. P. per hour and 1 lb. of coal will vaporize 8.3 lbs. of water, then 33.1 ÷ 8.3 = 4 lbs. coal required per boiler horse power per hour.

We find by consulting the table on page 210 that one pound of air at 70 degrees has a volume of 13.34 cu. ft. Good, safe practice will allow 19 lbs. of air per pound of coal consumed. (The theoretical requirement being 11.75 cu. ft.). Therefore, 19 lbs. of air at 70 degrees has a volume of 19 x 13.34, or 253.5 cu. ft. so 252.5 x 4 x 1,000 ÷ 60 = say 17,000, and thus 17,000 C. F. M. is required.

Thumb Rule for Required Air at any Temperature

C. F. M. = HxCxLxQ ÷ 60
in which:

H = Boiler Horse Power

C = 4 for good coal, 5 fair, 6 poor, and 7 very poor coal.

L = 19 (pounds air per pound coal)

Q = Volume of 1 lb. of air at temperature to be handled by fan from table page 210.

Having determined the volume required the necessary draft is then needed.

The following figures will be found to suit average conditions:

Forced Draft

Using Ordinary grates and with a Duct System—1.75" water gauge, static pressure.

Fan blowing into Ash Pit without Ducts —1.25" water gauge Static Pressure.

Using stokers with chain grates of the underfeed type, the pressure varies from 1.5" water gauge static pressure to 6" water gauge static pressure, depending on the type and make of stoker.

Induced Draft

Suction Pressure required for a system with boilers running at rated capacity:

1.0" W. G. Static Press. normal

1.25" W. G. Static Press. 25% overload

1.75" W. G. Static Press. 50%–60% overload

Formula for Determining Required Pressure

If it is desired to figure the total press-ure or draft required this may be found by the following formula:

$$P = (R \div K)^2$$

In which

P = Pressure in inches of water

R = Rate of combustion in pounds per sq. ft. of grate per hour.

K = Constant for different grades of coal

The values of K are as follows:

Bituminous Lump	34
Bituminous Run Of Mine	30
Bituminous Slack	26
Anthracite Pea	22
Anthracite Buckweat	20
Anthracite Birdseye	15
Anthracite Culm	10

The above values are based on a Coal consumption of 20 pounds per sq. ft. grate area per hour.

In case economizers are used with an installation, between 50 to 60% should be added to the static pressure.

Fig. No. 1, page 210, gives another method of determining the pressure required for efficient combustion, it being a chart showing the difference in pressure between the ash pit and the space over the fire. To this pressure must be added the loss due to the frictional resistance of the boiler breeching, economizers, etc.

Typical installations of Jeffrey Fans direct connected to Engine for supplying draft to boilers in Power Houses.

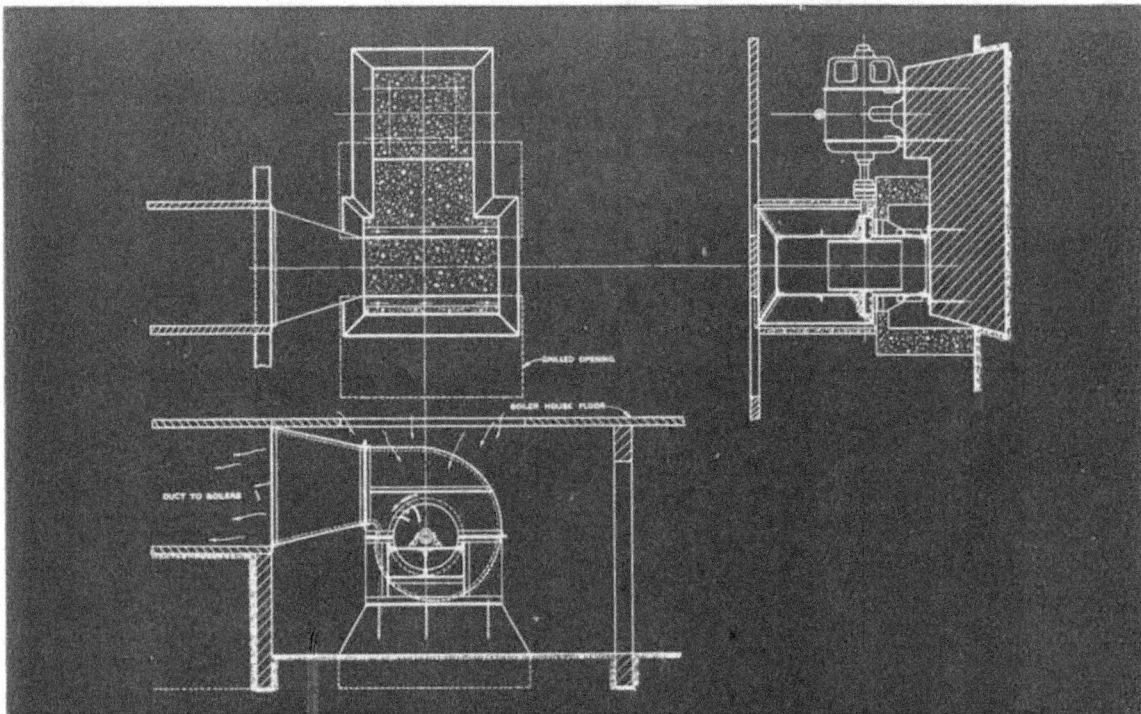

The above cut shows a 5'-6" x 2'-0" Jeffrey Fan direct connected to a variable speed motor, used for Forced Draft in connection with Underfeed Stokers.

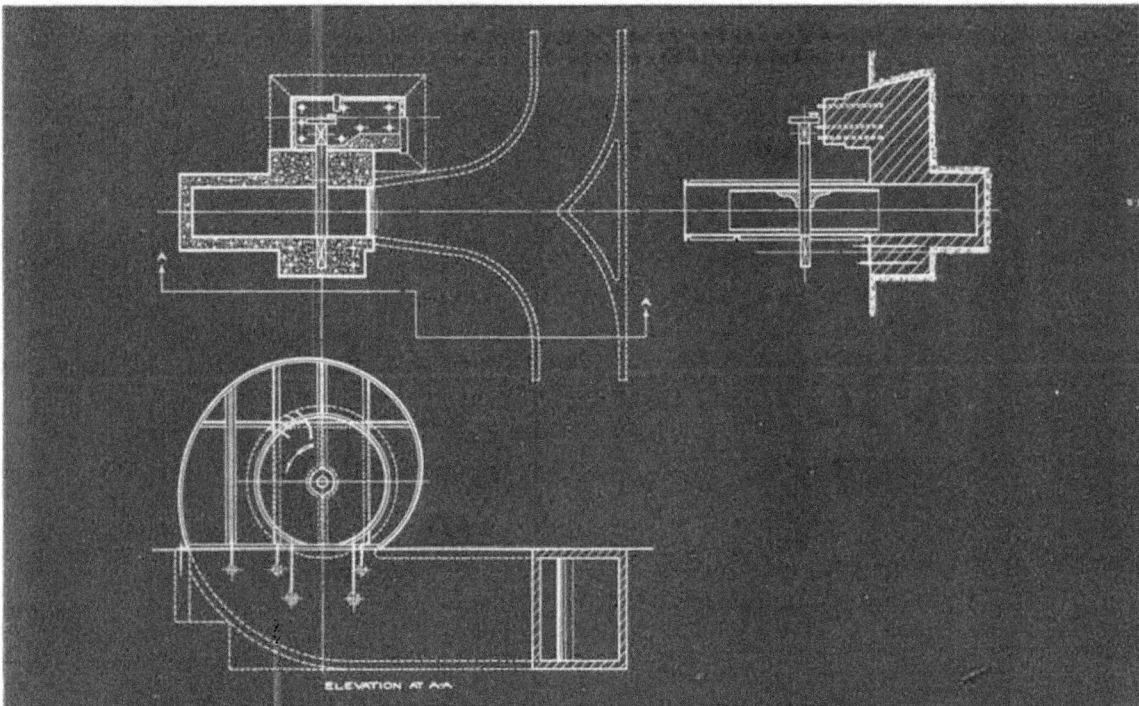

A 10'-0" x 2'-6" Single Inlet Jeffrey Fan direct connected to an engine. This is a typical arrangement popular with those who do not care for very high speeds for their Mechanical Draft Units.

A Double Installation of Jeffrey Fans at the Pennsylvania Coal Co. One of these fans supplies the air to the boilers, while the other is held as a reserve unit.

A Battery of Forced Draft Fans at the Pennsylvania Coal Co. One fan is always held in reserve in case of accident to the fan operating. The drawing at top of following page shows how they function.

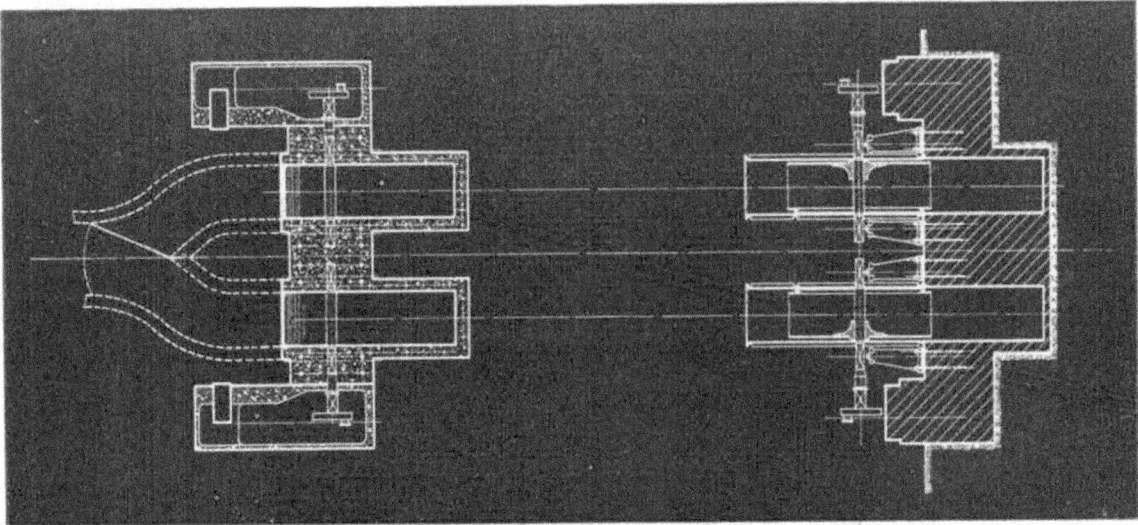

A pair of 10'-0" x 3'-0" Single Inlet Jeffrey Forced Draft Fans. One is held in reserve in case of emergency. Where plenty of room is available units may be placed side by side as shown.

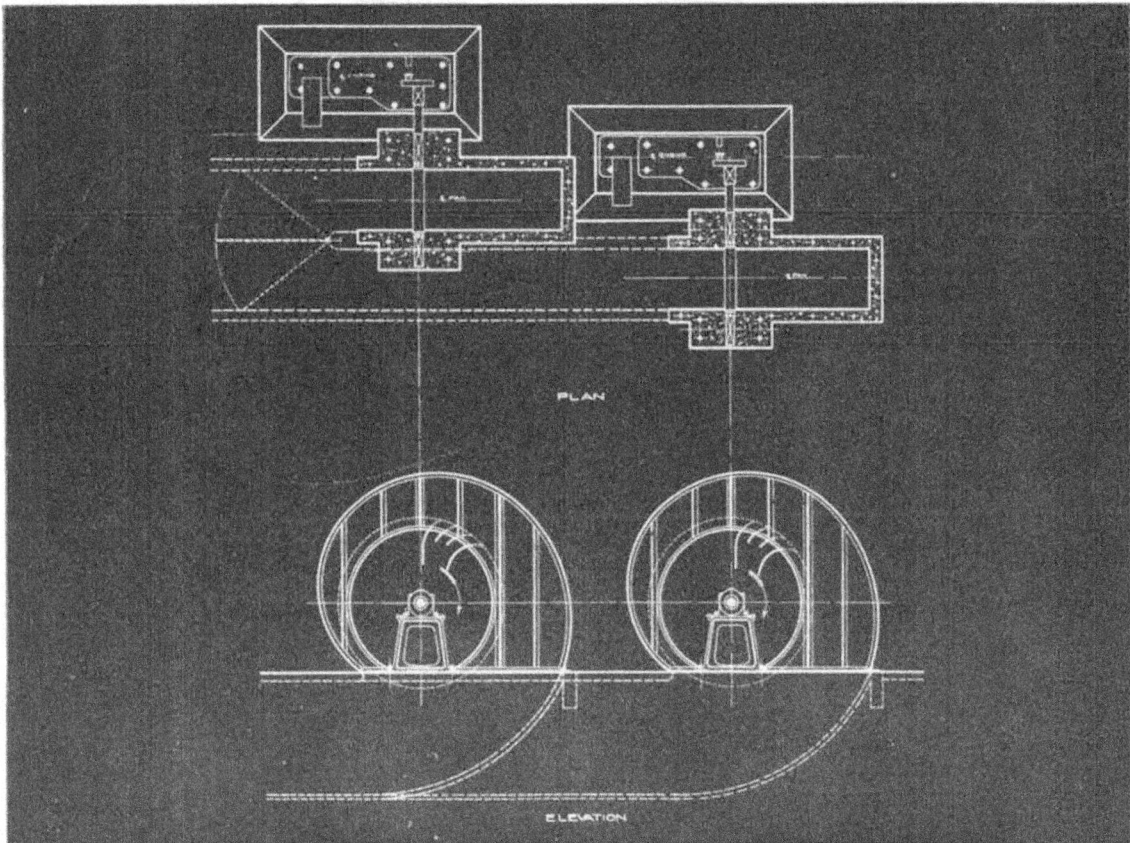

A pair of 12'-0" x 3'-6" Single Inlet Jeffrey Forced Draft Fans. In this case owing to lack of room the reserve unit is placed as shown.

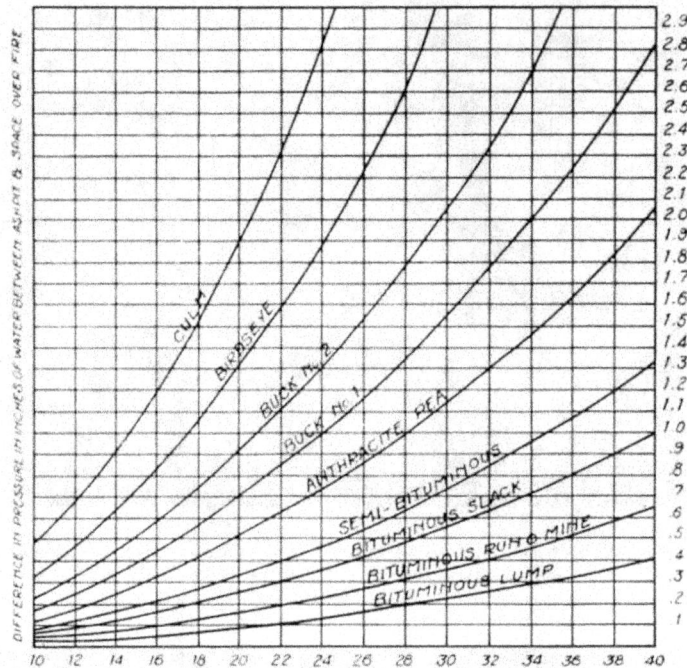

Figure 1
Coal consumption per sq. ft. of grate area per hour.

Volume and Density of Air at Various Temperatures

(At Atmospheric Pressure of 14.7 Lbs.)

Temperature Degrees	Volume of 1 Lb. of Air Cubic Feet	Density or Weight of 1 cu. ft of Air Lbs.	Temperature Degrees	Volume of 1 Lb. of Air Cubic Feet	Density or Weight of 1 cu. ft. of Air Lbs.
0	11.583	.086331	320	19.624	.050959
32	12.387	.080728	340	20.126	.049686
40	12.586	.079439	360	20.63	.048476
50	12.84	.077884	380	21.131	.047323
62	13.141	.076096	400	21.634	.046223
70	13.342	.07495	425	22.262	.04492
80	13.593	.073565	450	22.89	.043686
90	13.845	.07223	475	23.518	.04252
100	14.096	.070942	500	24.146	.041414
120	14.592	.0685	525	24.775	.040364
140	15.1	.066221	550	25.403	.039365
160	15.603	.064088	575	26.031	.038415
180	16.106	.06209	600	26.659	.03751
200	16.606	.06021	650	27.915	.035822
210	16.86	.059313	700	29.171	.03428
212	16.91	.059135	750	30.428	.032865
220	17.111	.058442	800	31.684	.031561
240	17.612	.056774	850	32.941	.030358
260	18.116	.0552	900	34.197	.029242
280	18.621	.05371	950	35.454	.028206
300	19.121	.052297	1000	36.811	.027241

Index

INDEX

A

	Page No.
Apron Conveyors, Steel	153 to 170

B

Battery Locomotives	201
Belt Conveyors	
Application	126, 152
Belting	142, 143
Carriers	134 to 136
Guide Pulleys	137
Idlers	138
Trippers	139, 140
Belt Conveyors, Portable	198, 199
Bin Valves	88 to 90
Bucket Elevators	63 to 70

C

Cable Drag Scraper Conveyor	196, 197
Car Unloaders, Portable	199
Carriers	
Belt	134 to 136
Pivoted Bucket	4 to 36
Century Rubber Belting	142
Coal Crushers	185 to 194
Coal Loaders and Unloaders, Portable	198 to 200
Coal Storage Equipment	195 to 200
Conveyors	
Apron	153 to 170
Belt	126 to 152
Drag Chain for Ashes	122 to 124
Pan	172, 173
Pivoted Bucket	4 to 36
Portable Belt	198, 199
Scraper	91 to 121
Spiral	174, 175
V-Bucket	37 to 62
Crushers	185 to 194

D

Depressed Pan Conveyor	172, 173
Drag Chain Ashes Conveyor	122 to 124
Drag Scraper, Cable	196, 197

E

Elevators	
Bucket	63 to 70
V-Bucket	37 to 62

F

Fans	
Industrial	203 to 210
Mechanical Draft	203 to 210
Feeders, Apron and Plate	81 to 87, 153 to 170

H

	Page No.
Hangers for Belt Conveyors	138
Helicoid Spiral Conveyor	174
Hoist, Skip	177 to 184
Hoppers, Track	79 to 87

I

Idlers for Belt Conveyor	138
Industrial Locomotives	201, 202
Industrial Fans	203 to 210

L

Larries, Weigh	71 to 78
Loaders for Coal	198 to 200
Locomotives, Industrial	201, 202

M

Maxlife Belting	143
Mechanical Draft Fans	203 to 210

P

Pan Conveyors	172, 173
Pivoted Bucket Carrier	4 to 36
Plate Feeders	81 to 85
Portable Belt Conveyors	198, 199
Portable Crusher	190
Portable Loaders	198 to 200

R

Radial Loader	200
Rubber Belting	142, 143

S

Scraper Conveyors	91 to 121
Cable Drag Scraper Conveyor	196, 197
Portable Scraper Loader	208
Single Roll Coal Crusher	185 to 194
Skip Hoist	177 to 184
Spiral Conveyors	174, 175
Steel Apron Conveyors	153 to 170
Storage Battery Locomotives	201

T

Track Hoppers	79 to 87
Trippers	
Belt Conveyor	139 to 140
Pivoted Bucket	21

U

Unloaders	
Car	199
Portable Belt	198, 199
Portable Scraper	200

V

V-Bucket Conveyors	37 to 62
Valves	
Bin	88 to 90
Scraper Conveyor Valves	96

W

Weigh Larries	71 to 78

THE following Catalogs, covering other types of Jeffrey Material Handling Equipment, will be mailed to you upon request:

106—Barrel Elevators.

175—Belt Conveyors for General Service.

231A—Storage Battery Locomotives.

244B—Bucket Elevators.

257—Scraper Conveyors.

258—Wood Apron Conveyors.

274—Portable Conveyors and Stackers.

368—Swing Hammer Pulverizers.

370—Industrial Fans.

386—Coal Mine and Tipple Equipments.

389—Pulp and Paper Mill Equipment.

397—Cement Mill Equipment.

398—Spiral Conveyors.

415—Chains, Attachments, Sprockets.

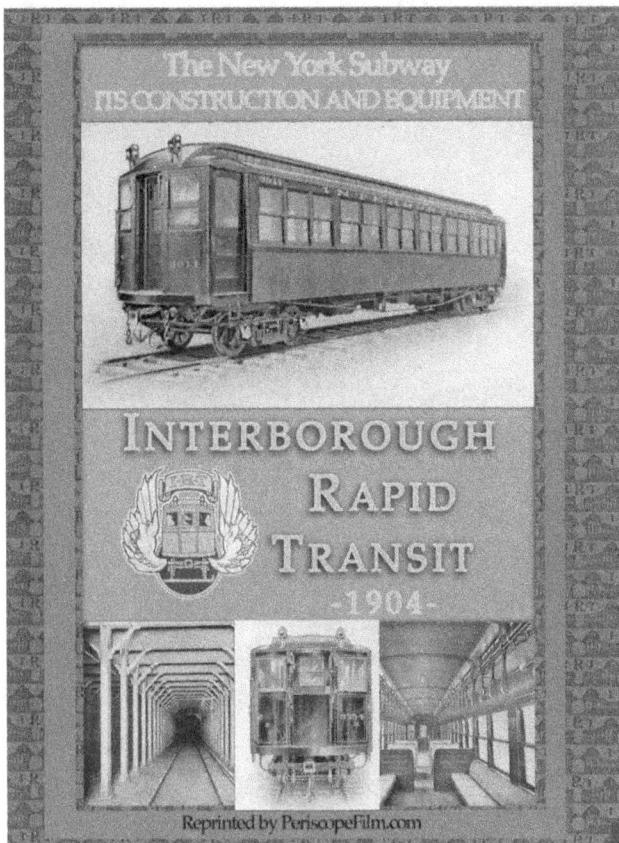

The New York Subway
ITS CONSTRUCTION AND EQUIPMENT

INTERBOROUGH
RAPID
TRANSIT
-1904-

Reprinted by PeriscopeFilm.com

On October 27, 1904, the Interborough Rapid Transit Company opened the first subway in New York City. Running between City Hall and 145th Street at Broadway, the line was greeted with enthusiasm and, in some circles, trepidation. Created under the supervision of Chief Engineer S.L.F. Deyo, the arrival of the IRT foreshadowed the end of the "elevated" transit era on the island of Manhattan. The subway proved such a success that the IRT Co. soon achieved a monopoly on New York public transit. In 1940 the IRT and its rival the BMT were taken over by the City of New York. Today, the IRT subway lines still exist, primarily in Manhattan where they are operated as the "A Division" of the subway. Reprinted here is a special book created by the IRT, recounting the design and construction of the fledgling subway system. Originally created in 1904, it presents the IRT story with a flourish, and with numerous fascinating illustrations and rare photographs.

Originally written in the late 1900's and then periodically revised, A History of the Baldwin Locomotive Works chronicles the origins and growth of one of America's greatest industrial-era corporations. Founded in the early 1830's by Philadelphia jeweler Matthais Baldwin, the company built a huge number of steam locomotives before ceasing production in 1949. These included the 4-4-0 American type, 2-8-2 Mikado and 2-8-0 Consolidation. Hit hard by the loss of the steam engine market, Baldwin soldiered on for a brief while, producing electric and diesel engines. General Electric's dominance of the market proved too much, and Baldwin finally closed its doors in 1956. By that time over 70,500 Baldwin locomotives had been produced. This high quality reprint of the official company history dates from 1920. The book has been slightly reformatted, but care has been taken to preserve the integrity of the text.

NOW AVAILABLE AT
WWW.PERISCOPEFILM.COM

A HISTORY OF THE
BALDWIN
LOCOMOTIVE
WORKS
1831-1920

Reprinted by PeriscopeFilm.com

ELECTRIC RAILWAY DICTIONARY

By Rodney Hitt

Associate Editor, Electric Railway Journal

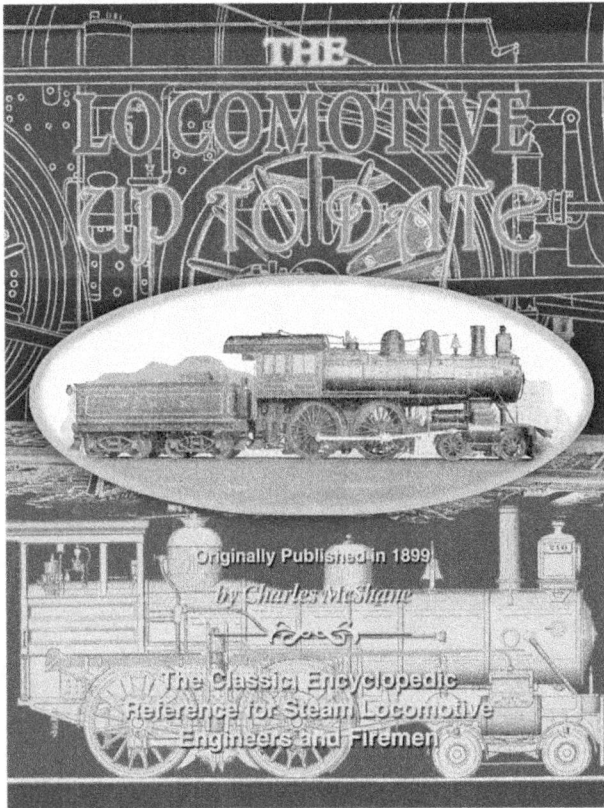
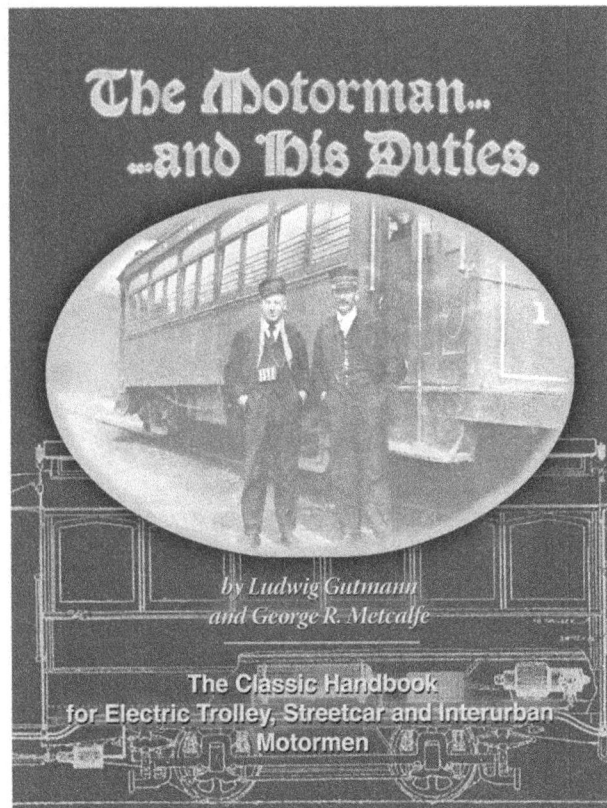

www.ingramcontent.com/pod-product-compliance
Lightning Source LLC
Chambersburg PA
CBHW081501200326
41518CB00015B/2343